D1237563

Autonomous Ground Vehicles

For a listing of recent titles in the
Artech House Intelligent Transportation Systems Series,
turn to the back of this book.

Autonomous Ground Vehicles

Ümit Özgüner
Tankut Acarman
Keith Redmill

ARTECH HOUSE

BOSTON | LONDON
artechhouse.com

Library of Congress Cataloging-in-Publication Data
A catalog record for this book is available from the U.S. Library of Congress.

British Library Cataloguing in Publication Data
A catalogue record for this book is available from the British Library.

ISBN-13: 978-1-60807-192-0

Cover design by Vicki Kane

© 2011 Artech House
685 Canton Street
Norwood, MA 06026

10 9 8 7 6 5 4 3 2 1

Contents

CHAPTER 8

Preface

This book is based on class notes for a senior/graduate-level course I, Ümit Özgüner, have been teaching at The Ohio State University for 8 years titled "Autonomy in Vehicles." Portions were also presented in lectures at short courses in different international venues.

The course, and thus the book, focuses on the understanding of autonomous vehicles and the technologies that aid in their development. Before we get to see fully autonomous vehicles on the roadway, we will encounter many of the constituent technologies in new cars. We therefore present a fairly comprehensive overview of the technologies and techniques that contribute to so-called "intelligent vehicles" as they go through the stages of having driver aids, semiautonomy, and finally reach full autonomy.

The book content relies heavily on our own experiences in developing a series of autonomous vehicles and participating in a series of international "demonstration" or "challenge" events. Although the book will explain in substantially more detail on developments at The Ohio State University, the reader is provided with the basic background and is encouraged to read and appreciate the work at other research institutions participating in the demonstration and challenges.

My coauthors and I would like to thank all the sponsors of the research and development activity reported here. First and foremost is The Ohio State University College of Engineering, directly and through the Transportation Research Endowment Program (TREP) and the OSU Center for Automotive Research (CAR); the OSU Electrical and Computer Engineering Department; and the Transportation Research Center (TRC) testing facility in East Liberty, Ohio. We would like to thank Honda R&D Americas, National Instruments, OKI, Oshkosh Truck Co., Denso, and, finally, NSF, which is presently supporting our work through the Cyber Physical Systems (CPS) program.

A substantial portion of the work reported here is due to the efforts of students working on projects. A few have to be mentioned by name: Scott Biddlestone, Dr. Qi Chen, Dr. Lina Fu, Dr. Cem Hatipoglu, Arda Kurt, Dr. Jai Lei, Dr. Yiting Liu, John Martin, Scott Schneider, Ashish B. Shah, Kevin Streib, and Dr. Hai Yu.

Finally, I need to thank my wife and colleague Professor Fusun Özgüner for all her contributions.

CHAPTER 1

Introduction

1.1 Background in Autonomy in Cars

We define "autonomy" in a car as *the car making driving decisions without intervention of a human*. As such, *autonomy* exists in various aspects of a car today: cruise control and antilock brake systems (ABS) are also prime examples of autonomous behavior. Some systems already existing in a few models—advanced cruise control, lane departure avoidance, obstacle avoidance systems—are all autonomous. Near-term developments that we anticipate, with some first appearing as warning devices, are intersection collision warning, lane change warning, backup parking, parallel parking aids, and bus precision docking. These capabilities show either autonomous behavior, or can be totally autonomous with the simple addition of actuation. Finally, truck convoys and driverless buses in enclosed areas have seen limited operation.

Studies in autonomous behavior for cars, concentrating on sensing, perception, and control, have been going on for a number of years. One can list a number of capabilities, beyond basic speed regulation, that are key to autonomous behavior. These will all affect the cars of the future:

- Car following/convoying;
- Lane tracking/lane change;
- Emergency stopping;
- Obstacle avoidance.

In what follows in this book, we shall be investigating all of these basic operations in detail, and show how they are also part of more complex systems.

In each one of the above operations the car is expected to do self-sensing (basically speed and acceleration), sensing with respect to some absolute coordinate system (usually using GPS, with possibly the help of a map data base), and sensing with respect to the immediate environment (with respect to lane markings, special indicators on the road, obstacles, other vehicles, and so forth). We shall also be reviewing the relevant technologies in this book.

1.2   Components of Autonomy

1.2.1   Sensors

One of the key components of an autonomous vehicle is the sensors (see Figure 1.1). The vehicle has to first have measurements related to its own state. Thus it needs to sense its speed, possibly through its wheel sensors, and also understand its direction of motion. The vehicle could use rate gyros and steering angle measurements as standard devices for this data. Data from GPS can also be used in this regard.

The second class of sensors is used by the car to find its position in relation to other vehicles and obstacles. Vision systems, radars, lidars, and ultrasonic sensors can be mentioned.

Finally, the vehicle needs to find its position with respect to the roadway. Different sensors have been used with or without infrastructure-based help. Vision systems, radar, magnetometers, and RF ID tags can be mentioned. GPS data has also been used with good digital maps.

1.2.2   Actuators

To be able to control the car in an autonomous fashion, it has to be interfaced to a computer. The specific control loops are speed control (throttle control), steering control, and brake control. Unless we are dealing with automatic transmission, the gear shift also has to be automated. A car in which all actuation in these loops is electrical is called a "drive-by-wire" vehicle.

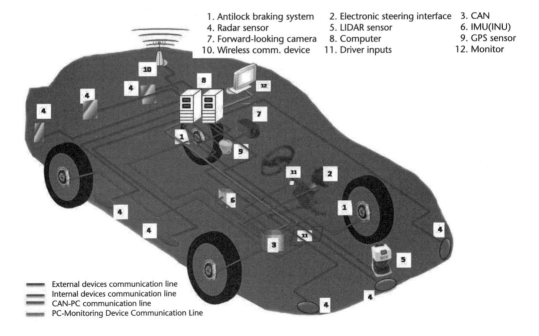

Figure 1.1   Generic concept of vehicle integrated by sensors and signal processor.

1.2.3   Communication

If a car has access to information through a wireless network, this too can help in the automation of driving. Communication can provide general information as well as real-time warnings to the driver data that can replace measurements.

One obvious use of communication between two vehicles is in exchanging data on relative location and relative speed. This can replace or augment data from sensors.

1.2.4   Intelligence

The intelligence in vehicles is introduced by the rules and procedures provided and programmed by humans. As such, as of the time of this writing, intelligent vehicles emulate the driving behavior of humans. The so-called intelligence is supplied by one or more processors on the vehicle. To this end we can classify three (sets) of processors: Those for handling sensor data, those for low level control algorithms and loop closure over the actuators, and finally those for establishing the autonomous behavior. We shall investigate their tasks in subsequent chapters.

1.3   Notes on Historical Development

1.3.1   Research and Experiments on Autonomous Vehicles

Most researchers in autonomous vehicles and AHS mention the 1939 Worlds Fair as the first example of exposition of a self-driving car. It was displayed to the public in the General Motors' "Futurama" exhibit. That concept car was illustrated by descriptions of cars automatically transporting people on convenient, stress-free trips on both urban streets and intercity highways. Technological development in support of this vision began in the late 1940s and early 1950s, with pioneering efforts by RCA Laboratories and GM Research. By 1960, GM had tested full-scale cars driving automatically on test tracks, with lane tracking, lane changing, and car following functions automated. The goal of these efforts was development of an automated highway system (AHS).

In the 1960s, one of the key efforts was the research and development effort under the leadership of Professor Robert Fenton and his team at The Ohio State University [1]. During these years vehicles were developed to test steering and speed controllers. Runs were accomplished at the Transportation Research Center (TRC) grounds, using active sensing (a live wire imbedded in the roadway) and with steering control done with analog feedback controllers. Autonomous runs were done with full speed control based on extensive studies of modeling of longitudinal motion and braking. An ingenious method was developed for providing reference speed profiles. Due to the lack of simple technology for real-time sensing, car following with real cars was difficult to perform. In Fenton's experimental setup, the distance to the car ahead was measured by using the physical connection with a spring-wound tape.

The contributions were not just related to individual self-driving cars. Advances were also made in establishing a hierarchical system monitoring and controlling all cars in a network of highways, fully occupied with automated cars, called intelligent vehicle highway systems.

In Japan, research on automated highway systems was started in the early 1960s in the Mechanical Engineering Laboratory (MEL) and the National Institute of Advanced Industrial Science and Technology (AIST). The automated driving system in the 1960s employed a combination of inductive cable embedded under a roadway surface and a pair of pickup coils at the front bumper of a vehicle for lateral control, and it was a cooperative system between a vehicle and infrastructure. In addition to the free agent, a rear end collision avoidance system between two automated vehicles based on roadway vehicle sensors was developed. The automated vehicle drove up to 100 km/h on a test track in 1967 [2]. A comprehensive test system for vision-based application development was built on the production-type vehicle in the late 1980s and stereo cameras were introduced to detect obstacles along lanes with curvatures and crossings at 50 km/h speed (see [3, 4]).

In Germany, in the beginning of the 1980s Professor Dickmanns and his team equipped a Mercedes-Benz van with cameras and other sensors. For safety reasons, initial experiments in Bavaria took place on streets without traffic. In 1986 the robot car "VaMoRs" from the same team managed to drive all by itself; by 1987 VaMoRs was able to travel at speeds up to 96 km/h, or roughly 60 mph. A number of these capabilities were also highlighted in the European PROMETHEUS Project (see, for example, [5]).

Starting with the PROMETHEUS Project, which involved more than 20 car manufacturers in Europe, some Italian research centers began their activity in the field of perception for vehicles. Among these, the CRF (FIAT Research Center) and the University of Parma developed some early prototypes, which demonstrated the viability of perception technologies like artificial vision.

In the United States, apart from the work at The Ohio State University, development of autonomous vehicles were undertaken in programs at Carnegie Mellon University, with links to robotics activity and vision-based expertise [6], and within the California PATH Program [7].

California Partners for Advanced Transit and Highways (California PATH) program is dedicated to increasing highway capacity and safety, and reducing traffic congestion, air pollution, and energy consumption by applying advanced technology. It funds the research projects selected among the proposals submitted throughout California and acts a collaboration between the California Department of Transportation (Caltrans), the University of California, other public and private academic institutions, and private industry (see http://www.path.berkeley.edu/). The previous PATH research projects elaborating *autonomy in vehicles* were the platoon control demonstration in 1997 in San Diego, California, where the autonomy was attributed with one vehicle capable of splitting, doing a lane change, falling back, and doing another lane change to join the platoon back autonomously. Advanced snowplow, advanced rotary plow, automated heavy vehicle control, and Bayesian automated taxi concerning vision-guided automated intelligent vehicles were the previous projects involving control and decision autonomy. Projects on sensing and sensor technology were directed towards inertial and GPS measurement on the vehicle for kinematic positioning and magnetic marker reference sensing on the road infrastructure to measure the relative vehicle position with respect to the center of the lane and to carry the roadway characteristics integrated in the form of the binary code into the north and south poles of those markers. Recent projects have been concentrated on transportation research for automated bus steering near bus stops in a city by using autonomous magnetic guidance technology and deployment of a research testbed to enable traveler communication through the wireless gateway. Choices about information type to be transferred to the traveler and the driver's communication handheld device are being determined by deploying a research testbed in the San Francisco Bay area.

In Europe, a project called "PReVENT" was integrated by the European automotive industry and the European commission to develop and demonstrate active safety system applications and technologies [8]. The main goals were to reduce half of the road accidents, render the European automotive industry more competitive, and disseminate transport safety initiatives. A public exhibition was organized on September 18–22, 2007, in Versailles, France, with many safety applications accompanied by safe speed, lateral support, intersection safety, collision mitigation, advanced driver assistance systems, and electronic stability control demonstrations, (PReVENT-IP project and its public deliverables are available at http://www.prevent-ip.org). Another current research activity on autonomous vehicle and demonstration has been developing in Europe under the canonic name "Cybercar." Cybercar has been introduced as the road vehicles with full automated driving capabilities based on a sensor suite, transportation system management, and also vehicle-to-vehicle, vehicle-to-infrastructure communication. The future concept of public transportation has been investigated since the early 1990s for carrying passengers or goods door-to-door through the network of roads and was implemented in 1997 to transport passenger at Schipol airport, the Netherlands. The Cybercar concept has been developed and expanded under a numerous number of European projects such as CyberCars, CyberMove, EDICT, Netmobil, CyberC3, and currently CyberCars-2 and CityMobil (for interested readers, please check the former Cybercar Web page available at http://www.cybercars.org/).

Through the years a number of developments have been introduced to the public in national and international challenges. We will summarize those below. Apart from those, two specific recent demonstrations, showing the activities of specific groups, are worth mentioning: A team led by Professor Alberto Broggi demonstrated a set of driverless vans as they went from Italy to China. The 8,000-mile journey ended in Shanghai, verifying the robustness of a number of technologies used for autonomous vehicles. A team supported by Google and led by Sebastian Thrun demonstrated a fully autonomous car in fairly dense urban California traffic, utilizing vision-based systems and Google Street View.

1.3.2   Autonomous Driving Demonstrations

Through the years, many demonstrations have been held showing the public the capabilities of autonomous vehicles, and by extension, underlining the sensor capabilities at the autonomous driving demonstrations. One of the most comprehensive highway-based demonstrations was held in 1997 on I-15 in San Diego, California, and showed the capabilities of cars, buses, and trucks in various automated highway scenarios. The demonstration (called Demo '97) was organized by the National Highway Systems Consortium. The key technologies were lane following using roadway-imbedded magnets, roadway-laid radar-reflective stripes (see Figure 1.2), or lane markers with vehicle-mounted cameras; and car following using laser or radar with or without the help of intervehicle communication. Scenario maintenance was accomplished by total preplanning and timing, by GPS- and map-based triggering, or by situation-based triggering [9].

Demo '97 was followed by a number of demonstrations around the world, notably in Japan, Holland, France, and the United States. In each case, not only were state-of-the-art technologies demonstrated, but the public was informed about what to expect in future cars.

In November 2000, the demonstration "Demo 2000 Cooperative Driving" was organized by the National Institute of Advanced Industrial Science and Technology in Tsukuba, Japan. Several platooning scenarios were accomplished by five automated vehicles in the enclosed test field. This platoon of five automated vehicles performed stop-and-go operation, platooning, splitting into two platoons demonstrating an exit ramp, merging into one platoon from two platoons imitating highway traffic at a ramp, and passing by the last vehicles. Furthermore, obstacle detection, platoon leaving, and joining tasks were accomplished. Autonomy was guaranteed by intervehicle communication. In addition, vehicles are equipped by differential global positioning system (DGPS) to assure the localization by the laser radar to detect obstacles and to measure the relative position between the vehicles. Vehicles of Demo 2000 were not communicating with infrastructure (i.e., road-to-vehicle communication) except for the DGPS. These automated vehicles were autonomously driven by exchanging information through intervehicle communication [10].

There were also a number of activities through the years supported by the Department of Defense on autonomy for off-road vehicles. In March 2004, the

Figure 1.2 Two autonomous cars from The Ohio State University team in Demo '97 performing a vehicle pass without infrastructure aids.

Defense Advanced Research Projects Agency (DARPA), to promote research in the area of autonomous vehicles, conducted a challenge for fully autonomous ground vehicles through a course between Los Angeles and Las Vegas. The winner had to have the fastest time and finish in less than 10 hours, without intervention. An award of $1 million was to be granted to the team that fielded the first vehicle to complete the designated route. The course was disclosed to entrants two hours before the race, in terms of over 2,000 waypoints. Nobody won. The challenge was repeated in October 2005 with a $2 million prize.

TerraMax '04 [shown in Figure 1.3(a)] was the intelligent off-road vehicle that our team, the *Desert Buckeyes* developed for Oshkosh Truck for the 2004 Grand Challenge. Of 106 applicants, it placed sixth [11]. ION [shown in Figure 1.3(b)] was the vehicle that the *Desert Buckeyes* outfitted for the 2005 Grand Challenge. As in 2004, drive-by-wire capability to control the steering, throttle, and brakes

(a)

(b)

Figure 1.3   (a) TerraMax before GC'04. (b) ION before GC'05.

was developed. Various sensors, including multiple digital, color cameras, LIDARS, radar, GPS, and inertial navigation units were mounted. We will review the Grand Challenge later in Chapter 4.

The 2007 DARPA Urban Challenge was performed in emulated city traffic and saw the return to urban vehicles. Autonomous ground vehicles competed to win the Urban Challenge 2007 prize by accomplishing an approximately 60-mile urban course area in less than 6 hours with moving traffic. The autonomous vehicle had to be built upon a full-size chassis compared to production-type vehicles and required a documented safety report. The autonomy capability was demonstrated in forms of a complex set of behaviors by obeying all traffic regulations while negotiating with other traffic and obstacles and merging into traffic. The urban driving environment challenged autonomous driving with some physical limits such as narrow lanes, sharp turns, and also by the daily difficulties in a daily urban driving such as congested intersections, obstacles, blocked streets, parked vehicles, pedestrians, and moving vehicles (http://www.darpa.mil/grandchallenge/overview. asp). Figure 1.4 shows the OSU car ACT that participated in the Urban Challenge, which was won by the Stanford team [12].

At the time of this writing, another challenge/demonstration was under preparation in the Netherlands: the Grand Cooperative Driving Challenge (GCDC), which plans to demonstrate the utilization of vehicle-to-vehicle (V2V) communication as it helps semiautonomous cars go through a series of collaborative exercises emulating cooperative driving in urban and highway settings (http://www.gcdc. net/). Team Mekar, which included Okan University, Istanbul Technical University, and The Ohio State University, joined the challenge with a V2V communication interface (Figure 1.5).

1.3.3 Recent Appearances in the Market

We are currently witnessing driver assistance systems with environmental sensing capabilities entering the automotive market. For many reasons (which include market acceptance and liability), those innovations are frequently first introduced into Japanese models before they are adopted in the European and American markets.

Starting with the Mitsubishi Diamante in 1995 and followed by Toyota in 1996, adaptive cruise control (ACC) systems employ automotive lidar or radar sensors to measure the distance, velocity, and heading angle of preceding vehicles. This information is used to improve on the longitudinal control of conventional cruise control systems. When a free roadway is detected, the system behaves just like a conventional cruise control. When a slower preceding vehicle is detected the ACC systems follows at a safe driving distance until the situation changes (e.g., due to a lane change maneuver of either vehicle). The system works well on highways or in similar operation conditions. It is designed as a comfort enhancing system (i.e., it neither primarily aims to nor is its sensor performance sufficient to provide safe longitudinal control by itself). The responsibility is kept with the driver and hence the operational domain is restricted (e.g. to a comfortable maximum deceleration of 2 m/s²). In European cars, the Mercedes S-Class introduced ACC (called Distronic) in 1999 followed by Jaguar's XKR the same year and BMW's 7er in early 2000. Despite initial and some remaining deficiencies, which mainly stem from the sensors, ACC systems have meanwhile become widely accepted in

Figure 1.4   OSU-ACT (Autonomous City Transport) before the 2006 DARPA Urban Grand Challenge.

upper-class vehicles. Function enhancements that allow seamless operation from highway cruising to stop-and-go traffic are under development.

Video sensors have first been introduced for lane departure warning (LDW) systems. These systems observe the white lane markings on the road and warn the driver when an unintended lane departure is detected. Market introduction has first focused on trucks such as the Mercedes or MAN flagships. In America the market was first approached by the small company AssistWare, which was later taken over by a large supplier.

Night vision systems assist the driver in bad visibility conditions such as night driving by displaying an image as observed by a sensor that can cope with such conditions. GM has first introduced night vision in its Lincoln Navigator employing a far infrared (FIR) spectrum. The Mercedes S-Class followed in 2005 with an alternative near-infrared (NIR) concept that employs a CMOS camera and NIR high beam headlamps.

In 2002 Honda introduced a lane-keeping assist system called the Honda intelligent driver support system (HIDS) in the Japanese market. It combined ACC with lane-keeping support. Based on lane detection by a video sensor, an auxiliary supportive momentum is added to the driver's action. With the collision mitigation system (CMS) Honda has followed an escalation strategy to mitigate rear-end collisions. When the vehicle approaches another vehicle in a way that requires driver action, initially a warning signal is activated when the driver does not react; a moderate braking is activated that increases to about 6 m/s^2 when a collision is unavoidable. Should the driver react but his or her reaction is insufficient to avoid the accident, the system enhances his or her action during all phases. Even though the system cannot avoid all accidents, the support of the driver and the speed reduction will reduce collision severity. Since 2005 an active brake assistance system that supports the driver to avoid or mitigate frontal collisions is available in the Mercedes S-class.

At the time of this writing, there was extensive activity in using wireless vehicle-to-vehicle communication technologies. We would expect these technologies to provide additional aid to the capabilities of autonomous vehicles.

1.4   Contents of This Book

The following chapters of this book will take the reader through different technologies and techniques employed in autonomous vehicle development.

Figure 1.5   Semiautonomous vehicle MEKAR, supported by Okan University, the Istanbul Technical University, and The Ohio State University, before the Grand Cooperative Driving Challenge.

Chapter 2 is an introductory chapter to provide an overview of the dynamics and control issues that need to be addressed in more detail later. It is a simple exposition especially useful for electrical and computer engineering students who have not thought about dynamics after their physics classes.

Chapter 3 introduces both a common architecture useful for most autonomous vehicles and a hybrid systems model as a common setting for designing and analyzing high-level control issues. The common architecture provides a setting to understand the software and hardware interactions. The hybrid system framework provides a unified approach to model the requisite "decision making" in the autonomous vehicle.

Chapter 4 introduces the different sensor technologies and their utilization, and discusses sensor fusion approaches for autonomous vehicles.

Chapter 5 discusses and develops the control issues first introduced in Chapter 2.

Chapter 6 provides an overview and introduction to maps, the use of maps and the related path planning issues.

Chapter 7 is an overview of vehicle-to-vehicle (V2V) and vehicle-to-infrastructure (V2I) communication issues and their relevance in autonomous vehicles and semiautonomous driving.

Chapter 8 provides a conclusion with some open problems.

Many examples and overviews of developed systems can be found in [1–12]. Throughout the text, examples are provided from the numerous vehicles we have developed and our experiences in participating in many national and international challenge and demonstration events.

References

[1]   Fenton, R. E., and R. J. Mayhan, "Automated Highway Studies at The Ohio State University—An Overview," *IEEE Transactions of Vehicular Technology*, Vol. 40, No. 1, 1991, pp.100–113.

[2]   Ohshima, Y., et al., "Control System for Automatic Driving," *Proc. IFAC Tokyo Symposium on Systems Engineering for Control System Design*, Tokyo, Japan, 1965.

[3]   Tsugawa, S., "Vision-Based Vehicles in Japan: Machine Vision Systems and Driving Control Systems," *IEEE Transactions on Industrial Electronics*, Vol. 41, No. 4, 1994, pp. 398–405.

[4]   Tsugawa, S., "A History of Automated Highway Systems in Japan and Future Issues," *2008 International Conference on Vehicular Electronics and Safety*, Columbus, OH, 2008.

[5]   Dickmanns, E. D., et al., "Recursive 3D Road and Relative Ego-State Recognition," *IEEE Trans. PAMI*, Vol. 14, No. 2, 1992, pp. 199–213.

[6]   Thorpe C., et al., *Vision and Navigation: The Carnegie Mellon Navlab*, Boston, MA: Kluwer Academic Publishers, 1990.

[7]   Chang, K. S., et al., "Automated Highway System Experiments in the PATH Program," *IVHS Journal*, Vol. 1, No. 1, 1993, pp. 63–87.

[8]   Schulze, M., "Contribution of PReVENT to Safe Cars of the Future," *13th ITS World Congress,* London, U.K., 2006.

[9]   Bishop, R., *Intelligent Vehicle Technology and Trends*, Norwood, MA: Artech House, 2005.

[10] Kato, S., et al., "Vehicle Control Algorithms for Cooperative Driving with Automated Vehicles and Intervehicle Communications," *IEEE Transactions on Intelligent Transportation Systems*, Vol. 3, No. 3, 2002, pp. 155–161.

[11]   Özgüner, U., K. Redmill, and A. Broggi, "Team TerraMax and the DARPA Grand Challenge: A General Overview," *2004 IEEE Intelligent Vehicle Symposium*, Parma, Italy, 2004.

[12]   Montemerlo, M., et al., "Junior: The Stanford Entry in the Urban Challenge," *Journal of Field Robotics*, Vol. 25, No. 9, September 2008, pp. 569–597.

The Role of Control in Autonomous Systems

Autonomy in vehicles requires control of motion with respect to the desired objectives and constraints. In this chapter, we would like to demonstrate basic lateral and longitudinal vehicle control. The driving, stopping scenario is first presented to illustrate the control effects of braking input applied to the fairly simple motion dynamics. Then, collision and obstacle ideas are briefly given and steering is presented as another control variable. Headway distance or relative distance between the follower and leader vehicle is used to illustrate the introduction of feedback to the vehicle longitudinal control. The perception-reaction time and the simple vehicle dynamics' responses are simulated along the commanded sequences. Turning around a corner and the effects of the longitudinal speed versus the lateral motion responses are added just after the cruise control scenario to illustrate the possible coordinated use of the inputs. Fully autonomous vehicles, path generation, and tracking are briefly presented as an introductory passage to the next chapters.

2.1 Feedback

2.1.1 Speed Control Using Point Mass and Force Input

Throughout this book we shall be using point mass models first. This will help to illustrate certain applications in a simple way. Although certain behavioral details are certainly lost with such models, it is also true that basic motions on a one- or two-lane highway, and relative motions between cars, can be more easily understood.

We first consider a vehicle model as a point mass subject to drive (traction) force, either to accelerate or to cruise at a constant speed, or subject to brake force to decelerate. Road friction is modeled by adding a damping coefficient to the point mass dynamics given by

$$m\ddot{x} + \alpha\dot{x} = f_d - f_b \tag{2.1}$$

where x denotes displacement of the vehicle, $\dot{x} = \dfrac{dx(t)}{dt}$, its velocity as a first-order time derivative of the displacement, and $\ddot{x} = \dfrac{d^2x(t)}{dt^2}$ denotes acceleration/deceleration of the point mass as a second-order time derivative of the displacement; m denotes the mass of the vehicle, α is the viscous friction coefficient modeling road friction, f_d represents drive force, and f_b represents brake force applied to the vehicle model. Simple analysis can show that a fixed driving force will provide a steady-state fixed velocity. Under the assumption that the initial conditions for the states are zero, at steady-state conditions, or more clearly when the vehicle model cruises at a constant speed with constant driving force, the change with respect to time is equal to 0, $\ddot{x} \equiv \dfrac{d^2x}{d^2t} = 0$, the point mass vehicle model dynamics become, $\alpha\dot{x} = f_d$ or $\dot{x} = \dfrac{f_d}{\alpha}$. This simple analysis clearly concludes that the velocity is proportional to the fixed driving force divided by the viscous friction coefficient. Obviously adjustment of the braking force f_b, especially proportional to the velocity, will give us the option of slowing down and stopping faster.

In the first simulation scenario, we consider the point mass model with road friction. Driving with constant speed followed by stopping with only road friction and mass of the vehicle model is simulated. The time responses of the model's velocity are plotted in Figure 2.1. The brake force is not applied during stopping or the decelerating time period; some amount of driving force is applied to drive the model with constant speed at the initial stage of the simulation scenario. The time-responses of the displacement are plotted in Figure 2.2.

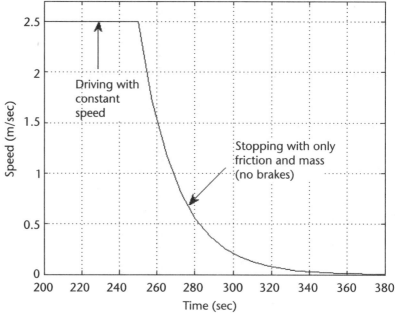

Figure 2.1 The time responses of vehicle model's velocity. The vehicle model is subject to road friction for stopping.

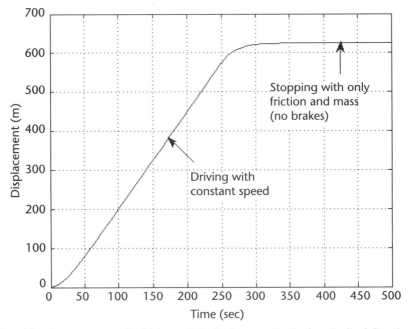

Figure 2.2   The time responses of vehicle model's displacement in the longitudinal direction subject to constant drive force and road friction.

2.1.2   Stopping

After considering the time responses of the displacement and the speed of the point mass vehicle model subject to road friction, we may consider stopping time and distance maintenance to consider time to hit or time to collide with the vehicle or the obstacle in front.

The braking force f_b may be changed to regulate the speed and the displacement of the vehicle model through the option of slowing down and stopping faster. Speed profile change versus increasing braking force is plotted in Figure 2.3, and the resulting displacement is given in Figure 2.4.

The presence of a potentially dangerous target can be determined by comparing the measured time to collision against a minimum time for stopping the vehicle safely. At any time t from the initial time t_0, the distance travelled can be written as [1]:

$$x(t) = x(t_0) + (t - t_0)\dot{x}(t_0) + (t - t_0)^2 \ddot{x}(t_0) \tag{2.2}$$

when collision occurs (i.e., for some time $t > t_0$), the distance $x(t) = 0$. Under the assumption that during braking maneuver for the time interval $[t_0, t]$, the speed and deceleration remain fixed at their initial values, which are denoted by $\dot{x}(t_0)$ and $\ddot{x}(t_0)$ at $t = t_0$, the time to collision (TTC) from the initial time t_0 to the instantaneous time t can be estimated as:

$$TTC = \frac{-\dot{x}(t_0) + \sqrt{\dot{x}(t_0)^2 - 4x(t_0)\ddot{x}(t_0)}}{2\ddot{x}(t_0)} \tag{2.3}$$

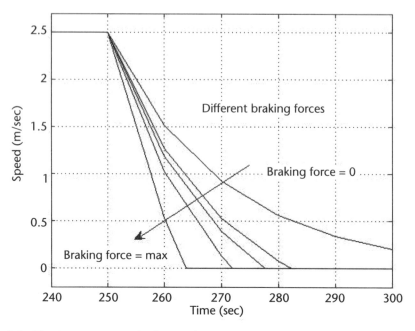

Figure 2.3   The time responses of velocity subject to different braking forces.

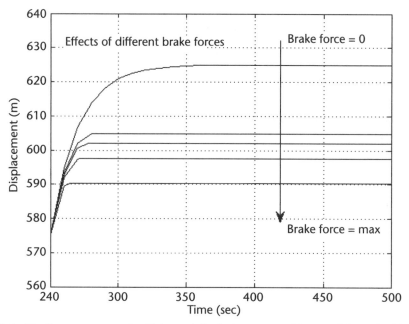

Figure 2.4   The time responses of vehicle model's displacement subject to different braking forces.

Minimum stopping time calculation with the assumption of a constant braking is derived by using time delay t_d, which includes processing and actuator delays; the brakes are applied with the maximum jerk J_{max}, which is already determined as the constant value for emergency maneuvering operations; the time to reach the maximum deceleration is denoted by a_{max}:

$$t_b = \frac{a_{max}}{J_{max}} + t_d$$

The value of t_{min} is calculated by using the condition given by

$$\dot{x}(t_d) + \int_{t_d}^{t_{min}} \ddot{x}(t)dt = 0 \tag{2.4}$$

Hence,

$$\dot{x}(t_d) - \frac{1}{2}J_{max}(t_b - t_d)^2 - a_{max}(t_{min} - t_b) = 0$$

$$t_{min} = \frac{\dot{x}(t_d) - \frac{1}{2}J_{max}(t_b - t_d)^2}{a_{max}} + t_b$$

It may be assumed that the service limits of 0.2g acceleration with a jerk 0.2 g.sec^{-1} is required for ride comfort. It is also assumed that emergency deceleration is about 0.4g. The United States standards on ride comfort of the passengers seated in the land vehicle were defined by the National Maglev Initiative Office [2]. However, these values are not necessarily fixed and they can be adjusted in accordance with the required control system performance.

2.1.3   Swerving

The capability to swerve around an obstacle can help indicate items such as maximum safe turn rate and required look-ahead distance to assure avoidance without rollover as illustrated in Figure 2.5 [3]. The look-ahead distance, denoted by D, can be generated using simple physics:

$$D = \sqrt{R^2 - (R - y)^2} + \dot{x}\, t_{react} + B \tag{2.5}$$

where $\dot{x}$ is the velocity in the longitudinal direction, and y is the displacement value that the vehicle has to travel in the lateral direction to arrive at the clearance point. The look-ahead distance is required to assure that the autonomous vehicle can accomplish the swerving task before closely approaching the obstacle or entering in the buffer space without rollover or sliding. Safe turn maneuver is calculated in terms of the maximum value of R_{min}, R_{roll}, or R_{slide} given by [4],

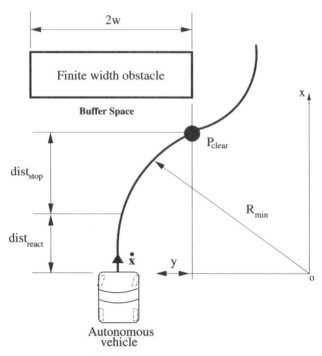

Figure 2.5 Swerving around an obstacle.

$$R_{roll} = \frac{2h\dot{x}^2}{wg}$$

$$R_{slide} = \frac{\dot{x}^2}{\mu g}$$

These equations involve many critical variables because of the large number of environmental and vehicle factors that effect the swerving abilities of a vehicle. The variables used in these equations are shown in Table 2.1.

2.2 A First Look at Autonomous Control

2.2.1 Car Following and Advanced Cruise Control

In the first simulation scenario, we present the initial investigations towards cruise control scenarios in which two identical vehicles are involved, the front one is called leading vehicle and the following vehicle is called subject vehicle for which the longitudinal motion is considered.

In this scenario, we consider maintaining headway with the car ahead traveling at a constant speed $\dot{x}_1$, and the following vehicle is traveling at $\dot{x}_2 \succ \dot{x}_1$. In this case, the following vehicle is going to hit the leading vehicle since its speed $\dot{x}_2$ and its drive (traction) force $f_d = C_2$ is higher than the leading vehicle's drive force $f_d = C_1$. Relative displacement or position between the leading and the following vehicle is plotted in Figure 2.6. In order to prevent collision with the vehicle in front,

Table 2.1   List of Typical Parameters to Determine the Swerving Abilities

h	The height of the center of gravity (cg).
t_{react}	Time required for the vehicle to react to an obstacle in sensor range.
$dist_{react}$	Distance in which the obstacle can be sensed and reacted to avoid.
$dist_{stop}$	Minimum distance necessary to stop the vehicle model before hitting the obstacle.
B	Additional buffer space included in calculations.
w	Half width of the obstacle.
R_{min}	Minimum turning radius of vehicle.
μ	Friction coefficient between the road and tire patches.

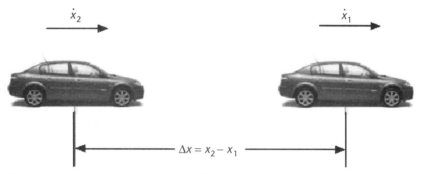

Figure 2.6   Cruise control scenario: leading vehicle and the cruise controlled follower vehicle.

the following vehicle has to maintain headway (a safe distance behind the leading vehicle). To maintain headway, the following vehicle's driver, going faster, must lay off the gas pedal and decelerate the vehicle with road friction. After settling the desired headway, the driver can apply some drive force to maintain headway. In Figure 2.7, the time responses of relative displacement are plotted; headway or relative displacement is decreased by the fast-going follower, and to prevent collision, at $t = 100$ seconds, traction force $f_d = C_2$ is commanded to be 0 (i.e., foot is taken off gas pedal), and the point mass model motion of the follower vehicle becomes: $m\ddot{x}_2 + a\dot{x}_2 = 0$. Speed is decreasing with the damping coefficient or road friction. At $t \cong 103.5$ seconds, to maintain headway settled, the same amount of traction force $f_d = C_1$ is applied to the following vehicle. To illustrate in detail the motion of relative speed and relative displacement change during the "foot off" and "constant headway" behavior, the magnification of the time interval is increased in Figures 2.8 and 2.9.

What if we have brakes? In the second headway simulation scenario, we investigate the use of brakes. In the first scenario, we have settled and maintained headway distance at approximately 9 meters. If we consider the use of brakes, as additional friction forces to road friction forces, we can get closer to the leading vehicle and apply brakes to avoid a collision and to increase headway (relative displacement) again.

We use the same amount of traction or drive forces as in the second simulation scenario, but we take the foot off the gas pedal at $t = 116$ seconds while decreasing displacement to the leading vehicle at 1.3 meters. As a human driver, first we take our foot off the gas pedal, and then we start applying brake force at $t = 117$

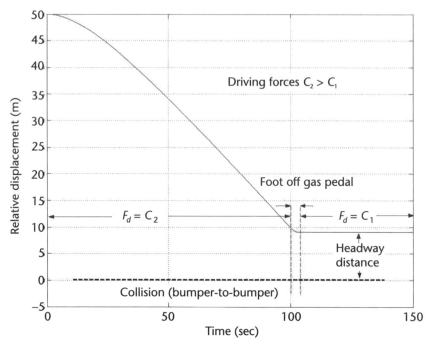

Figure 2.7 The time responses of relative displacement.

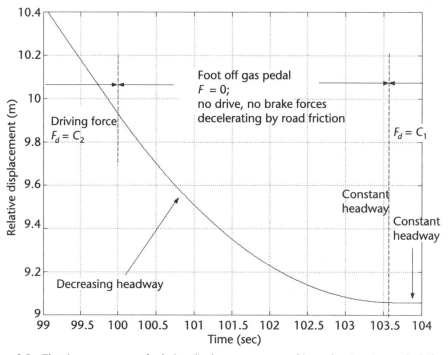

Figure 2.8 The time responses of relative displacement zoomed in at the time interval of "foot off gas pedal" and "constant headway."

seconds, as plotted in Figure 2.10. A 1-second delay is added to model the driver's foot displacement from the gas pedal to the brake pedal. Applying brake force until

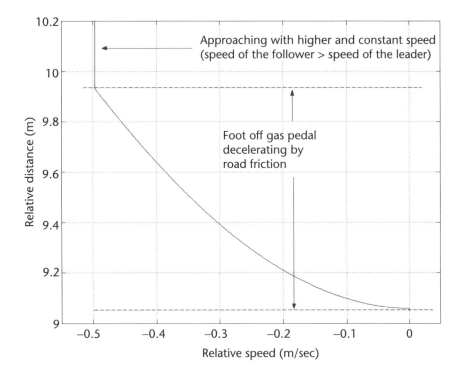

Figure 2.9   The time responses of relative displacement versus relative speed zoomed in at the time interval of "foot off gas pedal" and "constant headway."

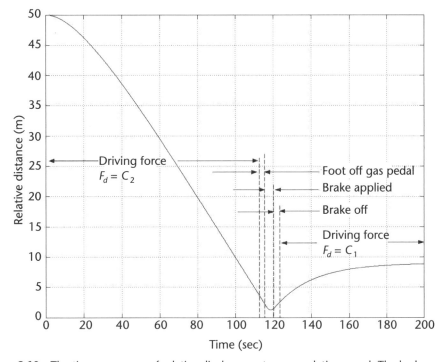

Figure 2.10   The time responses of relative displacement versus relative speed. The brakes are applied during the time interval $t = 117$ seconds and $t = 120$ seconds.

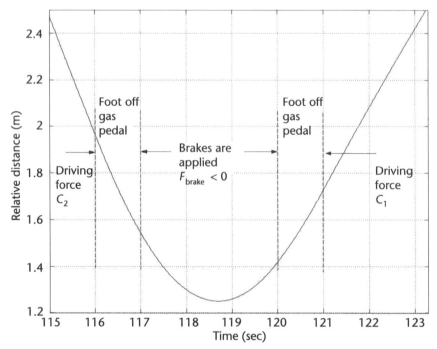

Figure 2.11   The time responses of relative displacement versus relative speed zoomed in the time interval of "foot off gas pedal" and "when brakes are applied."

$t = 120$ seconds, the relative position is increased as plotted in Figure 2.11. If brake forces are kept on, relative position continues to increase. Therefore, to maintain headway distance between two vehicles, drive force is applied again to the subject vehicle at $t = 121$ seconds at the same amount applied to the leading vehicle. Once again, a one-second time delay is presented to switch driver's foot from the brake pedal to the gas pedal.

In Figure 2.12, the time responses of relative displacement are compared for the cases of using additional friction brakes and the case of the road friction to maintain headway. In the case of braking, the subject vehicle can get closer because applied friction brake forces are decelerating the vehicle model motion faster in comparison to road friction force.

In Figure 2.13, the time responses of relative displacement versus relative speed are plotted. Applied brake forces are changing the sign of the relative speed between two vehicles. Approaching the leading vehicle, the follower subject to a higher drive force is decelerated by applying brake forces and the relative position or headway is increased. To maintain or settle the headway (i.e., at 9 meters as in both of headway maintenance scenario), the same amount of drive force is applied to stabilize the relative speed at zero (i.e., at the same speed).

2.2.2   Steering Control Using Point Mass Model: Open-Loop Commands

We consider the point mass model on the plane so as to present motion in the lateral and in the longitudinal direction,

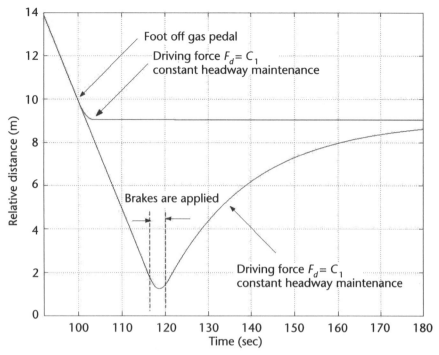

Figure 2.12   Comparisons of the scenarios when the brakes are applied and when the vehicle model motion dynamics are only affected by the road friction while maintaining larger headway.

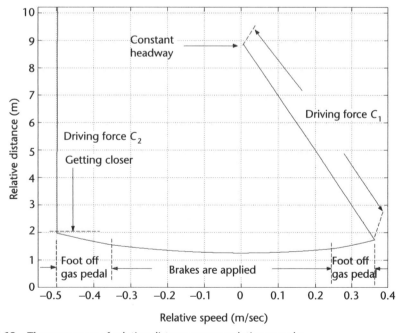

Figure 2.13   The responses of relative distance versus relative speed.

$$\begin{cases} \dot{x} = V_s \cos \Theta \\ \dot{y} = V_s \sin \Theta \\ \dot{\Theta} = u \end{cases} \qquad (2.6)$$

where x represents the position of the point mass vehicle model in the longitudinal direction, y represents the position of the point mass vehicle model in the lateral direction, Θ denotes the angle between the longitudinal and the lateral direction of the model, and u denotes the commanded steering wheel turn rate (rad/sec).

Note that this model assumes the speed to be constant. In fact, if the angle is zero, traveling along the x direction only, we will end up with a first-order differential equation in x, contrary to (2.1).

We consider fixed speed (V_s = constant). We simulate lane change maneuvering in Figure 2.14. The wheel steering angle is applied at $t = 5$ seconds and the point mass vehicle model's motion is plotted. Changing lanes is accomplished when the position of the vehicle model is shifted from the center of the left lane to the center of the right lane. Lane width is chosen to be 3.25 meters; therefore the left lane center is placed to be at 1.625 meters and the center of the right lane is placed to be at 4.875 meters as illustrated in Figure 2.15.

The wheel steering angle, Θ, and the wheel steering angular rate, u, are plotted in Figure 2.16 to accomplish lane change maneuvering. Double lane change maneuvering is accomplished by a right lane change operation followed by a left lane change operation (i.e., shifting the center of gravity of the vehicle model from the center of the left lane to the center of the right lane and vice versa). Displacement in both of the longitudinal and the lateral directions is plotted in Figure 2.17. The wheel steering angle, Θ, and wheel steering angular rate, u, are plotted in Figure 2.18 to accomplish double lane change maneuvering task.

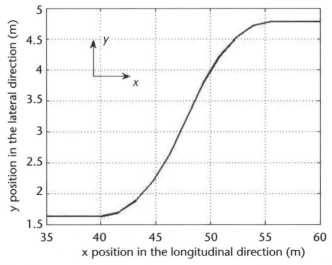

Figure 2.14 The time responses of lane change maneuvering. Displacement in the longitudinal direction is compared with displacement in the lateral direction.

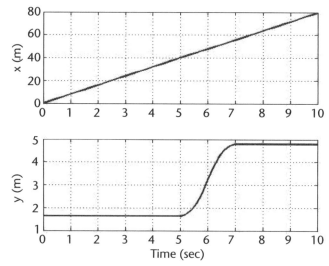

Figure 2.15   The time responses of displacement in the longitudinal and lateral direction.

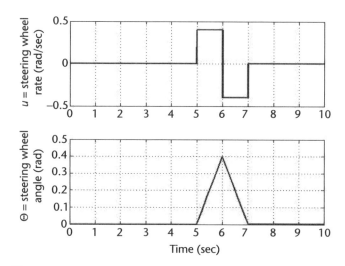

Figure 2.16   The time responses of steering wheel angle and steering wheel angle rate.

Both the lane change and the corner-turning operations are the ones that can be conceived to be open-loop—that is, preplanned operations. As in any system, the main distinction between open- and closed-loop is going to be related to the robustness. The open-loop examples provided work fine if certain assumptions are met exactly. For lane change, these could be assumptions regarding our car being exactly in the middle of the lane, the lane being a precise width that we know beforehand, the car responding exactly as its model predicts, that no other inputs exist, and so forth. Otherwise, we shall have to resort the closed-loop commands; to be able to generate those closed-loop commands we need to accomplish sensing and tracking the roadway. A preplanned operation illustration example is given next.

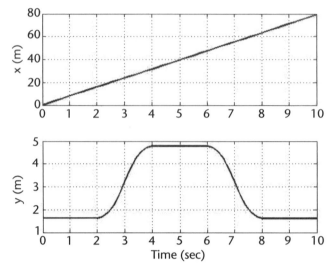

Figure 2.17 The time responses of displacement in both of the longitudinal and the lateral direction for double lane change maneuvering.

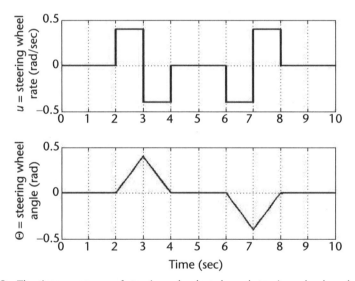

Figure 2.18 The time-responses of steering wheel angle and steering wheel angle rate.

Using the same point mass model on the plane, tracking the infrastructure is simulated by using the open-loop input steering wheel turn rate command or its sequences. The tracking scenario is illustrated by the road with a curvature at 30 meters in the longitudinal direction with $\frac{\pi}{4}$ (rad) angle. In the first scenario, the point mass model is steered to track the road at 30 meters in the longitudinal direction (i.e., at the moment the point mass displacement is at 30 meters and it is applied the open-loop steering wheel turn rate, u). Before the corner, the steering wheel turn rate is kept constant, $u = 0$, leading to displacement in the longitudinal direction.

The simulation scenario about turning case is repeated for $V_s = 10$ m/sec constant longitudinal speed and two different open-loop wheel turn rates. When the point mass model is steered by $u = 1$ rad/sec at the corner situated at 30 meters in the longitudinal direction, displacement of the point mass model in both of the longitudinal and lateral direction is not adequate to track the turning road (see Figure 2.19 for the tracking instance, when the curvature starts and 1 rad/sec turn rate is applied, the point mass model displacement in the longitudinal direction is 5 meters where the fictive lane boundary in the right side is exceeded). The steering angle of the point mass model (taking the time integral of the turn rate) is applied for the time interval [3; 4] seconds and it is plotted with a dash-dot line in the Figure 2.20.

To assure open-loop tracking, the turn rate command value may be increased. In the second case, $u = 2$ rad/sec for the time duration of $t = [3; 3.5]$ seconds (half-seconds shorter compared to the lower turn rate scenario) is applied at the corner; this steering angle waveform is plotted with solid line in Figure 2.20. The turning radius of the point mass maneuver may be reduced but the tracking maneuver is not successfully achieved due to the road lane departure on the turning side where it is supposed to track the lane boundary. Therefore, additional command sequencing may be necessary to avoid lane departure followed by turning. The turn rate is modified with respect to time as follows:

$$
u = \begin{cases}
2 & \text{during} & t = [3; 3.5] & \text{sec} \\
0 & \text{during} & t = [3.5; 3.7] & \text{sec} \\
-0,2 & \text{during} & t = [3.7; 4.75] & \text{sec} \\
0 & \text{during} & t = [4.75; \infty] & \text{sec}
\end{cases}
$$

Lane departure after turning may be avoided by the input turn rate command sequence plotted with dashed line in Figure 2.20.

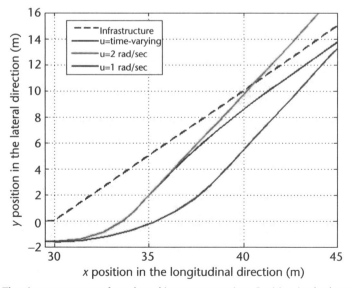

Figure 2.19   The time responses of road tracking maneuvering. Position in the longitudinal direction versus position in the lateral direction is plotted.

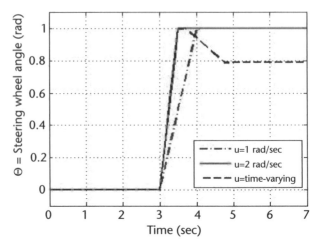

Figure 2.20 The waveform of the steering angle (its time derivative gives the turn rate).

2.2.3   Steering Control Using Point Mass Model: Closed-Loop Commands

We shall now consider steering with a closed-loop command, coming from sensors detecting our status with respect to the roadway.

It is assumed that the car is able to measure the lane deviation at some look-ahead distance away. We shall assume that this deviation is measured in terms of an angle ψ as illustrated in Figure 2.21. (We shall discuss the technology needed to make such a measurement in Chapter 4. At this point, let us assume such a measurement is possible, and that it is instantaneous and noise-free.)

Consider the basic equations for the point mass with fixed speed again:

$$\begin{cases} \dot{x} = V_s \cos \Theta \\ \dot{y} = V_s \sin \Theta \\ \quad \dot{\Theta} = u \end{cases} \tag{2.7}$$

We propose a steering wheel command

$$u \; = \; K\psi \tag{2.8}$$

Simulation to test such a lane tracking algorithm will rely on making a representation of the curved road. Such a representation can be done in terms of a data file (a point by point representation of the curve) or a polynomial or other closed-form expression.

We shall now move on to a very specific curve, namely, a 90° corner. Turning a corner is an important maneuvering task. Reference establishment is shown in Figure 2.22. A two-step reference trajectory is proposed to accomplish the corner turn.

The point mass vehicle, coming to the crossroad from the left side, is going to turn right. Reference establishment for turning right perpendicularly is constituted in two steps:

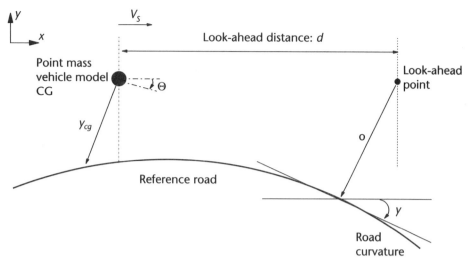

Figure 2.21   Tracking the road curvature.

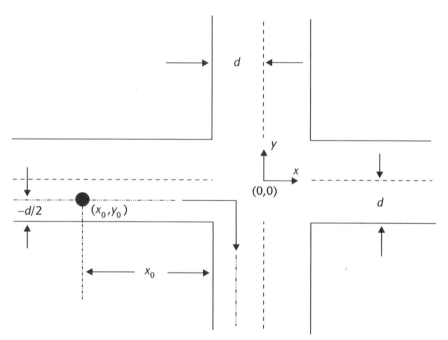

Figure 2.22   Turning corner and reference establishment.

1. *Lane keeping:* Coming onto the corner while maintaining the reference position in the lateral direction, denoted by y_0, and increasing the reference position in the longitudinal direction, denoted by x_0 in Figure 2.22; reference is established to approach to the corner:

$$x_0 \to -\frac{d}{2}$$

$$y_0 = -\frac{d}{2}$$

(2.9)

2. *Approaching the center of the corner:* Turning the corner, a new reference is established satisfying a fixed reference position in the longitudinal direction, denoted by $x_0 = -d/2$, and decreasing the reference position in the lateral direction, which is denoted by $y_0 = -d/2$, to the final reference value y_f.

$$x_0 = -\frac{d}{2}$$

$$y_0 = -\frac{d}{2} \to y_f$$

(2.10)

In Figure 2.23, the turning maneuver task is illustrated. In this plot, the initial reference for the lateral position is chosen constant and equal to the center of the lane, which is $y_0 = -1.625$ meters. Displacement in the longitudinal direction, denoted by x, is increasing while approaching to the corner. To achieve turning task around the corner, a second step of reference generation is established as constant longitudinal position, which is $x_0 = -1.625$ meters. New reference means a

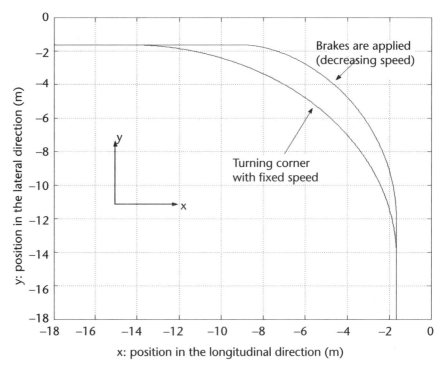

Figure 2.23 Comparison of displacement during corner maneuvering. Displacement in the longitudinal direction versus lateral direction is plotted for the case when speed is constant and when it is reduced.

perpendicular turn followed by displacement in the lateral direction to its final value denoted by y_f, which is a negative value smaller than the initial lateral position denoted by y_0. The vehicle displacement in the longitudinal direction is kept constant at $x_0 = -1.625$ meters by the imposed new reference set.

In Figure 2.23, turning task with constant speed is accomplished. Since the vehicle motion is subject to constant drive force, the turning radius of the maneuver is tending to be a large value around the corner. Speed can be reduced while approaching the corner and resumed to its fixed value after turning the corner. In Figure 2.24, the time responses of the turning task are compared for the case of constant and reduced speed while the steering wheel is kept the same for both of the cases where it is commanded to accomplish corner turning.

In practice, a trajectory is established and the vehicle follows a sequence of points along that trajectory. Two fundamental aspects of the autonomous driving problem are the relationship between a sequence of waypoints and the concept of lane boundaries and the controlled following of the sequence of points by the vehicle. A sketch of a vehicle on a path with waypoints and lanes is shown in Figure 2.25.

Lanes are obvious in highway systems and urban routes, whereas off-road driving presents a set of constraints subject to drivability of different areas and providing the possibility of lanes. Furthermore, the cars are subject to limited turning capabilities, and the continuous and smooth path can only be tracked by steering and driving. The generation of a smooth path is an important control design procedure. One of the basic approaches for a smooth path generation is polynomial tracking by sensing lane edge markers in a road infrastructure or continuous and

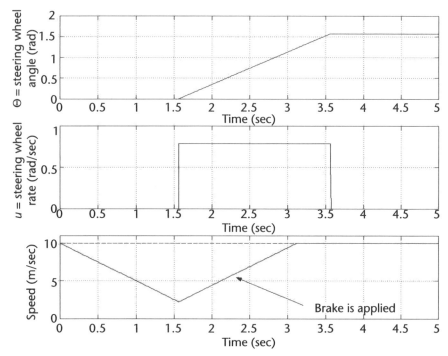

Figure 2.24 The time responses of the wheel angle and rate for turning corner maneuvering. The time responses of the speed are plotted for the case when speed is constant and when it is reduced to improve corner tracking.

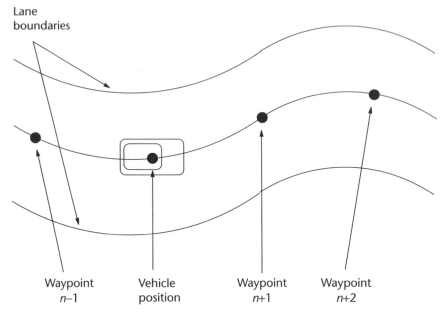

Figure 2.25 Roadway with waypoints and lane markings.

smooth reference generation, which connects the desired final point from the initial point with respect to some turning constraints.

2.2.4 Polynomial Tracking

Historically, in automated highway system studies, heading correction has been done based on lane detection. Lanes could be detected by vision technologies relying on lane edge markers [5], magnetic nails [6–8], radar reflective tapes [9, 10], or other such technologies where the infrastructure would support the sensing either actively or passively. There has been a preference to deal with so called clothoid models, in which the road curvature C, which is the inverse of the road's radius of curvature, is the basic parameter

$$C = \frac{1}{R} \tag{2.11}$$

and is assumed to be changed linearly with arc length

$$C = C_0 + \frac{dC}{dl} l = C_0 + C_1 l \tag{2.12}$$

Thus with constant speed and steering wheel turn rates, ideally there will be no deviation from the preferred path [11]. It can be shown that $C_1 = \frac{1}{A^2}$ is piecewise constant and A is the clothoid parameter. For small angular changes we can get an approximation,

$$Y = C_0 \frac{l^2}{2} + C_1 \frac{l^3}{6} \tag{2.13}$$

Again, for small deviations, this corresponds to a spline representation. This approach assumes the need for an independent representation of the roadway. Such a representation is not necessarily needed in implementation. Our activity in Demo '97 has been based on a look-ahead system simply measuring a heading error, and feeding back the error signal through a PID or P^2PID (squared error feedback) system without explicit modeling of the roadway in the controller.

2.2.5   Continuous and Smooth Trajectory Establishment

The cubic Bézier splines may be used to generate the continuous and smooth reference path to be tracked by the vehicles. The Bézier splines have been used extensively in the area of CAD, graphics, typography, and path planning. The Bézier splines are smooth and differentiable, and they are suitable for the path generation of limited turning radius vehicles.

The smooth and continuous path is generated by using Bézier splines. The four points P_1, P_2, P_3, and P_4 are defined; the Bézier curve always passes through the initial point P_1 and the destination point P_4. To assure correct heading angle (see also Section 5.4 for the requirements on the heading angle), the Bézier curve or the smooth path to be followed by the car is tangent to the line segment P_1P_2 at point P_1 and tangent to the line segment P_3P_4 at point P_4. The intermediate points P_2 and P_3 are used to adjust the slope.

A parametric equation for the Bézier curve can be obtained [12],

$$B_p(t) = At^3 + Bt^2 + Ct + P_1 \qquad 0 \le t \le 1 \tag{2.14}$$

$$A = P_4 - P_1 + 3P_2 + 3P_3 \tag{2.15}$$

$$B = 3(P_1 + P_3 - 2P_2) \tag{2.16}$$

$$C = 3(P_2 - P_1) \tag{2.17}$$

Considering the car maneuver control and a sequence of waypoints, the initial point P_1 may be defined by the current car position and the next point P_2 is chosen to be of distance D_1 in the direction of the heading from the current position, which is P_1. Towards the destination point P_4, the intermediate point P_3 is chosen to be of distance of D_2 while assuring an appropriate yaw angle. The offset distances D_1 and D_2, which may be tuned as the control points to generate a feasible trajectory, are plotted in the illustrative path generation example in Figure 2.26.

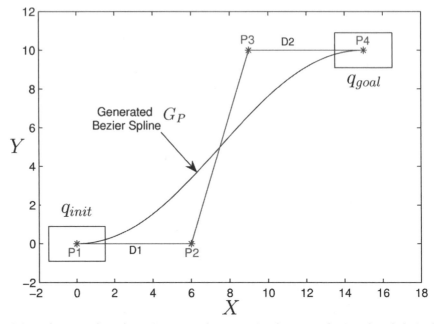

Figure 2.26 The smooth and continuous path generation between the initial and desired goal point by using Bézier curve techniques.

2.2.6 The Need for Command Sequencing

It should be clear that some way for "generating a command sequence" is needed, as indicated by the corner turning example, or earlier when doing lane changes. A lot can be accomplished by judicious use of reference signals while in lane-tracking mode. For example, a lane shift operation can be accomplished by inserting an additional bias signal, while the lane sensor simply switches to acquiring data from the next lane. But even then, a higher-level authority has to initiate the insertion of the bias signal at the appropriate "lane change" time. And in the corner-turning example, some sensor should be aware of the approach of the intersection, so as to initiate the reference-changing operation [13].

Command sequencing, as illustrated above, can be accomplished in one of two approaches, and possibly a combination of both. These are:

- Defining a finite set of states and transitioning through them. Each state leads to a set of feedback gains and/or reference signals. The model of the system is a combination of the vehicle dynamics jointly with a state machine, leading to a hybrid system.

- Defining a functional hierarchy, which under certain conditions again leads to a different set of feedback gains and reference signals. The hybrid system model is probably hidden, but still exists.

References

[1]   Ioannou, P. A., and C. C. Chien, "Autonomous Intelligent Cruise Control," *IEEE Transactions on Vehicular Technology*, Vol. 42, No. 4, 1993, pp. 657–672.

[2]   U.S. Department of Transportation, "Compendium of Executive Summaries from the Maglev System Concept Definition Final Reports," DOT/FRA/NMI-93/02, pp. 49–81.

[3]   Özgüner, Ü., C. Stiller, and K. Redmill, "Systems for Safety and Autonomous Behavior in Cars: The Darpa Grand Challenge Experience," *Proceedings of the IEEE*, Vol. 95, No. 2, 2007, pp. 397–412.

[4]   Robotic Systems Technology, "Demo III Experimental Unmanned Vehicle (XUV) Program; Autonomous Mobility Requirements Analysis," Revision I, 1998.

[5]   Redmill, K., "A Simple Vision System for Lane Keeping," *IEEE Intelligent Transportation Systems Conference*, Boston, MA, November 1997.

[6]   Zhang, W. -B., "National Automated Highway System Demonstration: A Platoon System," *IEEE Intelligent Transportation Systems Conference*, Boston, MA, November 1997.

[7]   Tan, H. -S., R. Rajamani, and W. -B. Zhang, "Demonstration of an Automated Highway Platoon System," *Proceedings of the American Control Conference*, Philadelphia, PA, June 1998.

[8]   Shladover, S., "PATH at 20—History and Major Milestones," *IEEE Transactions on Intelligent Transportation Systems*, Vol. 8, No. 4, December 2007, pp. 584–592.

[9]   Redmill, K., and Ü. Özgüner, "The Ohio State University Automated Highway System Demonstration Vehicle," *Journal of Passenger Cars*, SP-1332, Society of Automotive Engrs., 1999.

[10]   Farkas, D., et al., "Forward Looking Radar Navigation System for 1997 AHS Demonstration," *IEEE Conference on Intelligent Transportation Systems*, Boston, MA, November 1997.

[11]   Dickmanns, E. D., and B. D. Mysliwertz, "Recursive 3-D Road and Relative Ego-State Recognition," *IEEE Transactions on Pattern Analysis and Machine Intelligence*, Vol. 14, No. 2, February 1992, pp. 199–210.

[12]   Prautzsch, H., "Curve and Surface Fitting: An Introduction," *SIAM, Society for Industrial and Applied Mathematics Review*, Vol. 31, No. 1, 1989, pp. 155–157.

[13]   Özgüner, Ü., and K. Redmill, "Sensing, Control, and System Integration for Autonomous Vehicles: A Series of Challenges," *SICE Journal of Control, Measurement, and System Integration*, Vol. 1, No. 2, 2008, pp. 129–136.

CHAPTER 3

System Architecture and Hybrid System Modeling

3.1 System Architecture

3.1.1 Architectures Within Autonomous Vehicles

A generic structural/functional architecture for an autonomous vehicle is given in Figure 3.1 [1]. Each block is, of course, tailored to the required functionality and degree of autonomy of the vehicle. However, even for the simplest autonomous vehicle scenario (for example, a vehicle capable of driving autonomous in the highly controlled environment of an interstate highway), each of these subsystems would be required. The detailed architecture of such a vehicle (using the one we developed for Demo '97 as an example) is shown in Figure 3.2.

Figures 3.2 and 3.3 show the details of the architecture for two different autonomous vehicles that we developed in 1996 and 2007, more than 10 years apart. Although some technologies have changed, and in spite of one being for AHS and the other for autonomous urban driving, one can note the similarities between the two configurations.

Note that the Demo '97 car does not have a GPS system and relies totally on infrastructure-based cues to find its position with respect to the roadway [2]. The car has both a special (stereo) radar system that senses the radar-reflective stripe it straddles on the lane, and a vision system that senses the white lane markers on both sides of the lane. Sensing of other cars on the roadway is accomplished with the radar and a separate LIDAR unit.

In the Urban Challenge car developed by OSU (shown in Figure 3.3), the overall architecture is very similar [3]. No direct sensor-based lane detection was implemented, although the sensors utilized would certainly have accomplished it. The vehicle (based on the DARPA definition of the Challenge) simply used high-precision GPS signals.

3.1.2 Task Hierarchies for Autonomous Vehicles

In this section we shall outline the functionality covered by modules in our structural/functional architecture shown in Figure 3.1.

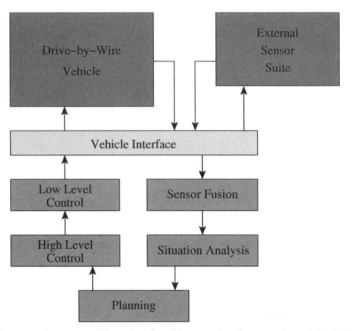

Figure 3.1   The generic structural/functional architecture implemented on OSU ACT.

3.1.2.1   High-Level Control

High-level control can be thought of as behavior generation, and in our autonomous vehicles is expressed as a hierarchy of finite-state machines. At the top level, overall families of behavior appear as independent state machines, which we designate as metastates.

The DARPA Grand Challenges of 2004 and 2005 were both off-road races. As such, the only behavior and thus the only metastate required would be path following with obstacle avoidance from point A to point B. However, since there was no path or lane that could be discerned from a roadway, the only method of navigation is to rely on GPS- and INS-based vehicle localization and a series of predefined waypoints. Obstacle avoidance techniques were needed, although in the less structured off-road scenario greater freedom of movement and deviations from the defined path were allowed. The Grand Challenge race rules ensured that there were no moving obstacles and different vehicles would not encounter each other in motion. General off-road driving would of course not have this constraint.

Fully autonomous urban driving introduces a significant number of new metastates—situations where different behavior and different classes of decisions need to be made. Figure 3.4 shows the highest-level metastate machine that defined the urban driving behavior of OSU-ACT. The DARPA Urban Challenge, although quite complex, did have fairly low speed limits, careful drivers, and no traffic lights. Visual lane markings were unreliable and thus true to life. The terrain was fairly flat, although some areas were unpaved, generating an unusual amount of dust and creating problems for some sensors.

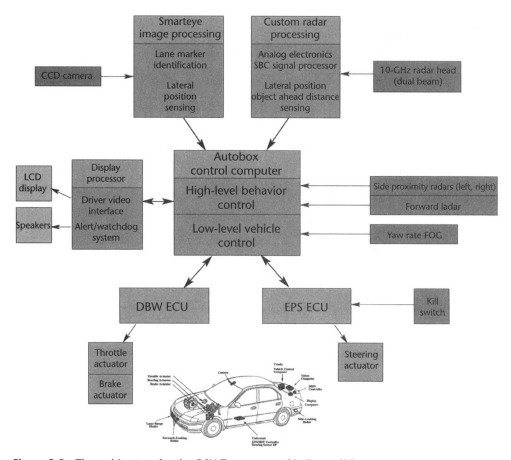

Figure 3.2 The architecture for the OSU Team car used in Demo '97.

3.1.2.2 Planning

Planning and path generation can take a number of forms depending on the required application. Between the starting and stopping locations, the vehicle was free to travel anywhere in the defined corridor. There were no high-level decisions to be made. The desired path could be fitting smooth functions through the established waypoints, and deviations from this desired path were generated as a reaction to local sensing information.

For the Urban Challenge, however, the behavior was defined as an ordered series of goal locations the vehicle was to attain, starting from any location, and the route had to be planned in real time over a map database defining a road network as well as parking lots (zones) and parking spaces. For this task, there were often multiple possible routes and an optimal route had to be identified based on estimates of travel time. The planning software also required the capability to remember blocked or impassible roads so that, if an initial plan failed, a new plan could be identified.

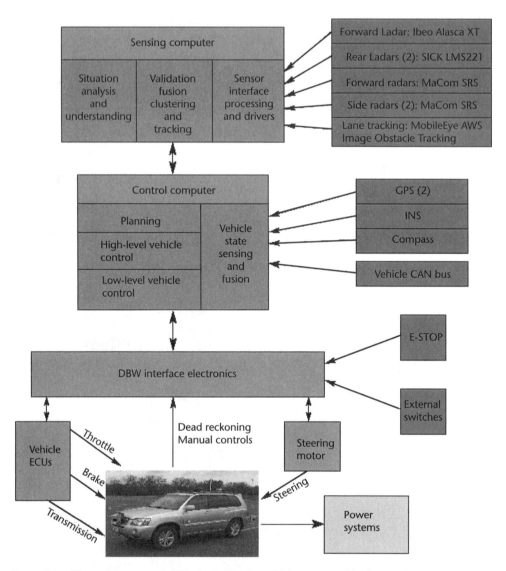

Figure 3.3   The architecture of ACT, the OSU urban driving car used in the DARPA UC.

3.1.2.3   Sensors and Sensor Fusion

We provide a brief introduction to sensor fusion as a portion of system architecture here. Further details are given in Chapter 4.

Vehicle Localization
A key element of autonomous vehicle technology is vehicle localization. All aspects of the system, from sensor processing and fusion to navigation and behavioral decision making to low-level lateral and longitudinal control, require accurate vehicle position, velocity, and vehicle heading, pitch, and roll information at a fairly high update rate. Providing this information requires the use of multiple sensors, including multiple Global Positioning System (GPS) receivers augmented with wide-area differential corrections for redundancy, inertial measurement units (IMU), and dead

reckoning sensors (wheel speeds, transmission gear and speeds, throttle, brake, and steering wheel position) provided on the vehicle, and a validation system to eliminate sensor errors, especially GPS-related step-change events caused by changes in differential correction status or the visible satellite constellation. To account for sensor errors, noise, and the different update rates of each sensor, an extended Kalman filter is applied to generate the required state measurements.

External Environment Sensing

There are some distinctions in considering sensing requirements for urban versus off-road applications. We list some noteworthy items here:

- For off-road applications, compensation for vibration and other vertical and rolling motions needs to be done in software or hardware, for example using the IMU and sensor data to specifically generate a ground plane that can be referenced while doing sensor validation and fusion. Sensor adjustments are also required to deal with dust, rain, and changing lighting conditions.

- For domains where there are many moving obstacles (i.e., urban applications), one may need to track individual obstacles at all times.

- Specific operations (parking, dealing with intersections, entering/exiting highways, and so forth) may use totally separate sensing and sensor architectures tailored to the task.

3.1.2.4   Sensor Interpretation and Situation Analysis

The term *situation* is defined to be knowledge concerning the vehicle and/or the prevailing scenario and surroundings. From a practical viewpoint, situations are the switching conditions among metastates and all the substates inside the high-level control state machines. Thus, the aim of situation analysis is to provide the high-level controller with all the switching conditions in a timely manner. The situation analysis software analyzes the current vehicle state, the current and upcoming required behavior for the route plan, the map database, and the sensor data to identify specific situations and conditions that are relevant to the vehicle's immediate and planned behavior.

For the off-road Grand Challenge scenario, the situation is always obstacle and collision avoidance. The software is required to identify, when analyzing the occupancy grid map blocks, the desired path and to adjust the planned path as needed.

For an urban scenario, we are interested in all the targets in our path and the targets in surrounding lanes or on roads intersecting our lane. We are not interested in targets that do not affect the current situation and planned behavior. While an autonomous vehicle is navigating through the city, many different situations may arise. The situations may vary if the vehicle is on a one-lane road, a two-lane road, an intersection, and so on.

Particularly critical for an autonomous vehicle are those situations related to intersections. When a car is approaching an intersection, it must give precedence to other vehicles already stopped. If the vehicles are stationary for a long time, the car must decide whether those vehicles are showing indecisive behavior. Other situations may involve road blockage in which the vehicle might carefully perform a U-turn, park in a parking space, and deal with dangerous behavior from other vehicles. All these situations must be identified and evaluated, and the resulting conclusions transmitted to the high-level controller in order for the vehicle to operate properly.

The path planning software provides information related to the current optimal path plan. Starting from the path, the situation analyzer can identify the location of the road and build a road model constructed from polygons derived from a spline curve fitting the waypoints defining the road shape. Such a road model design is particularly suitable for both accuracy and implementation purposes. In order to reduce computational costs and complexity, only the situations related to the current metastate or substates, as provided by the high-level control software, are checked.

3.1.2.5 Low-Level Control

Command Interface
In a two-level control hierarchy as shown in Figure 3.4, the low-level control receives operational instructions from the high-level control module. These instructions take the form of:

1. A path to be followed, defined by a set of approximately evenly spaced control points;
2. A desired speed;
3. Commands to indicate starting and stopping;
4. Special commands indicating motions, which can be fully implemented at the lower level.

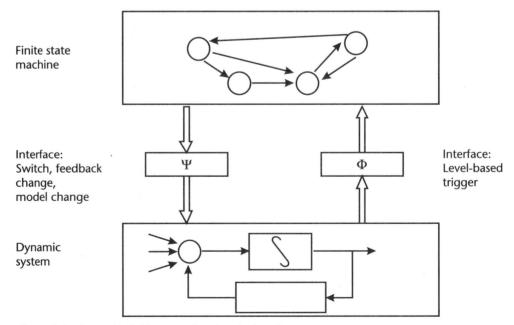

Figure 3.4 General hybrid system showing the interfaces.

Examples could be short movements along constant radius arcs, precision stops, and so forth. The low-level control will execute a given command set until the command is completed and the vehicle is in a stationary state, the vehicle has driven off the end of the path provided, at which point the vehicle will be stopped, or it receives a new command set.

Longitudinal Control

The interface and control of vehicle actuation is achieved by having a drive-by-wire car. Our experience has been that a simple control algorithm, for example a set of PID controllers, is adequate to generate a virtual torque command to achieve the commanded speed, and a state machine is used to select between the use of throttle, active braking, or engine idle braking. Speed commands are modified to constrain the acceleration and jerk of the vehicle to preset comfortable limits. There may also be emergency deceleration modes that are less comfortable.

Urban driving, in contrast to highway or off-road driving, requires the vehicle to execute a precise stop at predefined locations, for example the stop line of an intersection. To accomplish this, the low-level control determines the distance from the vehicle's current position to a line drawn through the specified stopping point and perpendicular to the vehicle's path of travel, taking into consideration the distance from the front bumper of the vehicle to its centroid. The speed of the vehicle is controlled to follow a specified, possibly nonlinear, deceleration trajectory.

Lateral Control

The path that the vehicle is to follow is specified as a set of control points. The lateral controller identifies both the current location of the vehicle and the look-ahead point (a prespecified distance ahead of the vehicle along its lateral axis) and extracts a subset of control points closest to each location. Constant radius circles are fitted to the points in each subset and these circles are used to compute the vehicle offset distances from the path and to estimate desired yaw rates. Each subset of points also defines a desired yaw angle for the vehicle. The offset distances, yaw angle error measurements, and desired yaw rates can be used to generating a feedback signal for the steering controller. There are a number of algorithms that can be used in this control loop, and a simple PID controller with fixed gains is not enough to cover all possible driving and path-shape scenarios. The variations here are speed dependent and turn-radius dependent.

3.2 Hybrid System Formulation

3.2.1 Discrete Event Systems, Finite State Machines, and Hybrid Systems

The high-level aspects of an intelligent vehicle can be modeled as a discrete event system. In this section we develop a modeling approach for modeling the discrete event system (DES). We represent the DES with a finite state machine (FSM), and then couple the FSM with a continuous time, dynamic system to create a hybrid system. One does not always need to go through the full formal development introduced here. Indeed, in many cases it is quite possible to directly develop an FSM.

We claim that hybrid system models are particularly useful in representing, simulating, and analyzing autonomous vehicles as they perform in complex environments, under prespecified scenarios and in possibly unplanned situations [4].

Let X denote the set of DES *states* and E denote the set of *events*. We define an *enable function*

$$g : X \to P(E) - \{\varnothing\}$$

which specifies which events are enabled at time t, and $P(E)$ denotes the power set of E. The DES state transition function is given by a set of operators

$$f_E ; X \to X$$

where E is a subset of g. The transition function specifies the next state when the event(s) in E occur.

Alternatively, the state transitions can be shown on a graph, where the nodes represent the states and the directed arcs are labeled with the individual events e in E, and are pointing from x to $f_E(x)$.

Now let us consider the case where the events e can be either generated externally, or can be generated by a separate continuous time, dynamic system. The interface from the continuous time system to the DES is described by a function Φ. Similarly, the DES also affects the continuous time system through a function Ψ (see Figure 3.4). Further details and examples in a nonvehicle context, can be found in [5].

In the following sections, we shall first look at ACC to formulate it as a DES and model it in terms of a finite state machine. We shall than consider first an obstacle avoidance operation and then a special bus service situation as hybrid systems.

3.2.2 Another Look at ACC

Consider first a cruise control system. The states of this system are given as {Cruise, Manual, Speed up}. It is assumed that the car slows to a speed below the desired cruise speed when it is in manual. When a return to cruise speed is desired, it can be accomplished either manually, or by setting cruise ON again. The events then are {Set ON, Set OFF}. It is assumed that the desired cruise speed is known.

A proposed state machine representing this DES is given in Figure 3.5. Students are urged to discuss if this machine is correct, and if, in fact, the above discussion was truly representative of what a cruise control system is expected to accomplish.

An advanced cruise control system is shown in Figure 3.6. It is of course assumed that our car has the means of detecting a car ahead of us and measuring its distance and relative speed. It will then slow down and match speeds, and will speed up if the car ahead speeds up. This speeding up process will continue until the car ahead gets to be faster than our desired cruise speed. At that time, we return to regular cruise operation.

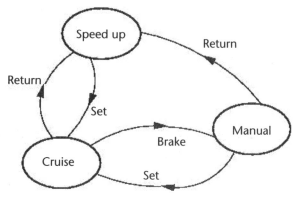

Figure 3.5   Standard cruise control logic (state machine model).

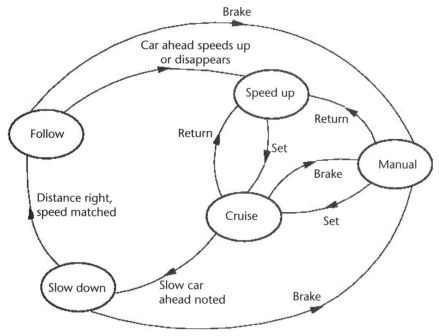

Figure 3.6   The state machine for an ACC system.

3.2.3   Application to Obstacle Avoidance

In this section we consider obstacle avoidance with the following assumptions.

1. The road is a multiple-lane, structured road, with detectable lane markers.
2. The car that we are driving always runs in the right lane if there is no obstacle ahead.
3. When an obstacle is detected ahead, the obstacle avoidance task is accomplished by a series of actions of changing to the left lane, passing the obstacle, and changing back to the right lane.
4. During the obstacle avoidance process, no other vehicles or obstacles will appear in the scenario. For example, the left lane is always empty so a left lane change is always risk-free.

Here we call an object "obstacle" based on its speed on the road. If an object has a speed less than certain threshold, v_{min}, our car will regard it as an obstacle and try to avoid it; otherwise, the car will just follow it.

A complete obstacle avoidance scenario is shown in Figure 3.7. As this figure illustrates, the whole obstacle avoidance procedure is divided into five stages as follows:

1. In this stage, we assume that there is no object ahead (within distance d_0). Thus, car 1 runs along the right lane of a two-lane road at speed v_1. Whenever an object is found within distance d_0, it enters stage 2.
2. In this stage, car 1 checks the speed of the object to see whether it is an obstacle or not. At the same time, car 1 still keeps running on the right lane, but it may slow down little by little as it is approaching the object. When the distance between car 1 and the object decreases down to d_1, it will either enter stage 3 if the detected object is considered as an obstacle, or just follow the object ahead, which will lead car 1 into another stage that is not shown in this scenario.
3. In this stage, car 1 turns left and changes to the left lane if it considers the object ahead as an obstacle. When the left lane changing is finished, car 1 enters stage 4.
4. In this stage, car 1 runs on the left lane until it has totally passed the obstacle. Then car 1 enters stage 5.
5. In this stage, car 1 turns right and changes back to the right lane. After that, car 1 switches back to stage 1.

Based on the assumptions and analysis above, we can design the obstacle avoidance system as shown in Figure 3.8. Obviously, the system will be a hybrid system. The continuous time system represents the acceleration system of car 1, which can switch among several dynamic models according to the car 1's current state. These models include standard (normal) running, following, approaching (object observed), left lane changing, passing, and right lane changing. The state switching

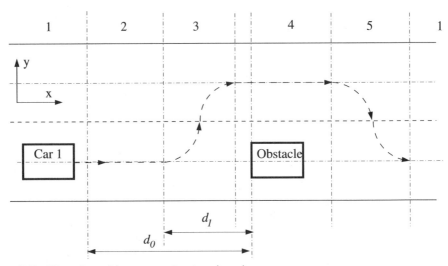

Figure 3.7   Obstacle avoidance on a structured road.

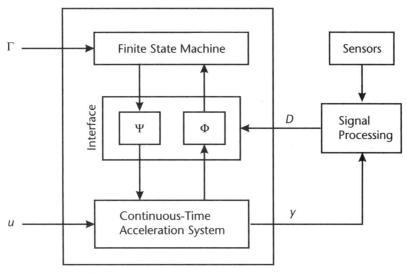

Figure 3.8   System model.

is controlled by a finite state machine. The interface provides the communication between the finite state machine and the continuous time system, which consists of two functions, ψ and ϕ. ψ is used to translate the state information to switching signals and send it to the acceleration system. ϕ is in charge of generating the events for the finite state machine according to the outputs of the system (y) and the data from the sensor system (D) based on some thresholds. Γ is the external control event, which can reset the state of the finite state machine. These control events are shown in Table 3.1.

Several sensor systems are used in the system:

1. A vision system based on cameras in the front to detect the lane markers, which is used to get the longitude position of the car on the road;
2. An obstacle detection system based on radar in the front to detect the objects ahead, which is used to get the distances and velocities of the objects ahead;
3. A side-looking radar system to on each side of car 1, which is used to check if car 1 has passed the obstacle.

The outline of the algorithm is represented by a finite state machine in the following figure. States "normal," "obstacle observed," "right lane change," "passing," and "left lane change," given in Figure 3.9, correspond to the stages outlined in Figure 3.7, respectively. The "follow" state is for the case when there is a low-speed vehicle ahead, but it is not considered as an obstacle. In this case, car 1 is

Table 3.1   External Control Events

Γ	*Meaning*
C1	Go to standard running state
C2	To make a left lane change
C3	To make a right lane change

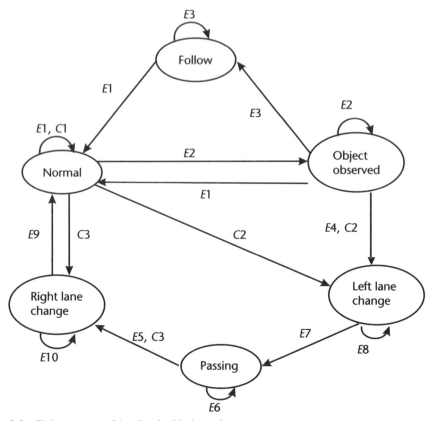

Figure 3.9   Finite state machine for double lane change.

assumed to be controlled by its ACC system and following the slow car ahead. Table 3.2 explains the events in the finite state machine, as well as the interface conditions to generate the events. Table 3.3 lists the parameters and variables used in the system.

As mentioned before, the interface part includes two functions, ψ and ϕ. The details of the thresholds specifying ϕ have been listed in Table 3.2. The function ψ is shown in Table 3.4.

Table 3.2   Events

Events	Interface Conditions to Generate the Events (ϕ)
E1	$d >= d_0$ or $v_2 >= v_{max}$
E2	$d_1 <= d < d_0$ and $v_2 < v_{max}$
E3	$d < d_1$ and $v_{min} <= v_2 < v_{max}$
E4	$d < d_1$ and $v_2 < v_{min}$
E5	$Pass = \text{true}$
E6	$Pass = \text{false}$
E7	$LLCF = \text{true}$
E8	$LLCF = \text{false}$
E9	$RLCF = \text{true}$
E10	$RLCF = \text{false}$

Table 3.3   Variables and Parameters

Constant Parameters	Meaning
d_0	The distance that an object ahead starts to be observed.
d_1	The distance that the object avoidance procedure starts.
v_{min}	The minimum lateral velocity of car 1 for following.
v_{max}	The maximum velocity of car 1.
Variables	Meaning
D	The distance between car 1 and the object ahead.
v_2	The velocity (lateral only) of the object ahead.
v_1	The lateral velocity of car 1.
Pass	Passing process indicator.
LLCF	Left lane change finished tag.
RLCF	Right lane change finished tag.

Table 3.4   Interfacing Function ψ

S: State of the Finite State Machine	$\psi(S)$: Model Switches
1: Normal	1: Standard running
2: Object observed	2: Approaching
3: Follow	3: Following
4: Left lane change	4: Left lane changing
5: Pass	5: Passing
6: Left lane change	6: Left lane changing

3.2.4   Another Example: Two Buses in a Single Lane

A single automated bus was set to carry passengers along a single lane from A to D and back. Due to increase in passenger count, the owners decided to have two automated buses. Instead of adding a full new lane, the buses were allowed to use the lane simultaneously, but have a passing lane/interchange developed at the approximate midpoint.

In this problem, we consider a bus system (shown in Figure 3.10), which consists of two stops (A, D), and two switching stations at B and C. Two buses will be running on the route. We are asked to design a control system for this scenario, so that two buses can run freely on the route without any collisions. The design is also expected to be extended to all the collision avoidance control system for future expansions.

The whole bus routing system can be treated as a hybrid system, which can be abstracted as given in Figure 3.11.

The two switches can be represented by two discrete time systems, which are actually two finite state machines (Figure 3.12). Each train is also a hybrid system, which includes a sensor system, a discrete time system, a continuous time system, and interfacing. The states and events of the switch system are listed in Tables 3.5 and 3.6. The bus system is a hybrid system itself, whose structure is outlined in Figure 3.13.

The dynamics of the continuous time subsystem S is simplified by the following equation:

Figure 3.10   The map for the bus.

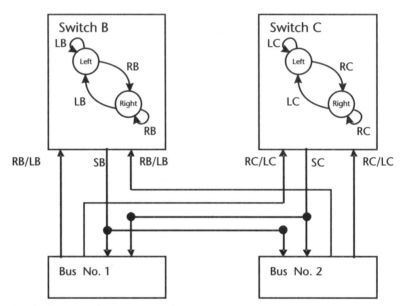

Figure 3.11   The bus interchange system diagram.

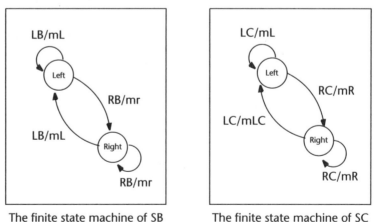

The finite state machine of SB   The finite state machine of SC

Figure 3.12   SB and SC finite state machines.

$$F = m\ddot{x} + \alpha\dot{x}$$

where F is the force.

Table 3.5   The States in the Switch System

State	Description
SB	Switch at point B, which can be either left or right.
SC	Switch at point C, which can be either left or right.

Table 3.6   The Events in the Switch System

Events	Description
LB	Switch at point B is to connect section AB with section BC_2 (input from the buses).
RB	Switch at point B is to connect section AB with section BC_1 (input from the buses).
LC	Switch at point C is to connect section CD with section BC_1 (input from the buses).
RC	Switch at point C is to connect section CD with section BC_2 (input from the buses).
mLB	Switch at point B is to connect section AB with section BC_2 (manual command input).
mRB	Switch at point B is to connect section AB with section BC_1 (manual command input).
mLC	Switch at point C is to connect section CD with section BC_1 (manual command input).
mRC	Switch at point C is to connect section CD with section BC_2 (manual command input).
SB	Output event, indicate the state of switch B.
SC	Output event, indicate the state of switch C.

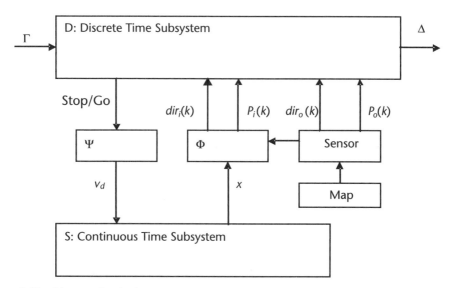

Figure 3.13   Diagram for the bus system.

If we further design an automatic cruise control system for the bus, we can get the following:

$$F = k_1(v_d - \dot{x}) + k_2 \int (v_d - \dot{x})dt + F_0$$

Thus the dynamics of the system is determined by the desired velocity v_d only. Then, the continuous time subsystem will simply work at one of the two models:

1. *Acceleration/maintain optimal speed: $v_d = v_0$;*
2. *Deceleration/stop: $v_d = 0$.*

Based on the idea above, the continuous time subsystem can be designed as in Figure 3.14.

The discrete time subsystem for each bus consists of two states: one for the position of the bus on the map, and the other for the dynamic status of the bus, as shown in Tables 3.7–3.9. Each state is running in a finite state machine, as shown in Figures 3.15 and 3.16.

The events "stop" and "go" here are generalized events for collision avoidance, which is generated based on the output of both the continuous time system (x) and the output of the sensor system (P_o, dir_o). In our problem these events are stimulated according to the truth table in Table 3.10. Only the conditions for Stop$_i$ to be

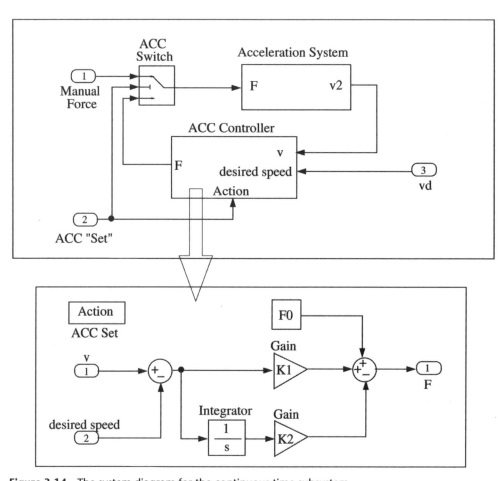

Figure 3.14 The system diagram for the continuous time subsystem.

Table 3.7 States in the Discrete Time Subsystem

State	Description
P_i	The position (in terms of the sections) of bus i on the map, which can be AB, BC$_1$, BC$_2$, or CD.
V_i	The desired dynamic status of bus i, which can be either acceleration or braking.

Table 3.8   Description of the States

State	Description
SB	Status of switch B imputed from switch B.
SC	Status of switch B imputed from switch C.
XA_o	$x_o = x_A$, (or $|x_o - x_A| < \varepsilon$), means the detected bus o arrives stop A.
XAB_o	$x_A < x_o < x_B$, means the detected bus o is between stop A and B.
XB_i	$x_o = x_B$, (or $|x_o - x_B| < \varepsilon$), means the detected bus o reaches point B.
XBC_o	$x_B < x_o < x_C$, means the detected bus o is between points B and C.
XC_o	$x_o = x_C$, (or $|x_o - x_C| < \varepsilon$), means the detected bus o reaches point C.
XCD_o	$x_C < x_o < x_D$, means the detected bus o is between point C and stop D.
XD_o	$x_o = x_D$, (or $|x_o - x_D| < \varepsilon$), means the detected bus o arrives stop D.
Forward$_o$	$dir_o = 1$, means the detected bus o is if moving from left to right.
Stop$_i$	Bus i is decelerating/braking to stop.
Go$_i$	Bus i is accelerating to/running at its desired speed.

Table 3.9   The Internal Events/Input/Output of the Discrete Time Subsystem

State	Description
SB	Status of switch B imputed from switch B.
SC	Status of switch B imputed from switch C.
P_o	Status of the obstacle (another bus that is detected by the sensor), can be AB, BC$_1$, BC$_2$, or CD.
dir_o	Moving direction of the obstacle from the sensor system.
XA_i	$x_i = x_A$, (or $|x_i - x_A| < \varepsilon$), means bus i is arrives stop A.
XAB_i	$x_A < x_i < x_B$, means bus i is between stop A and B.
XB_i	$x_i = x_B$, (or $|x_i - x_B| < \varepsilon$), means bus i is arrives at point B.
XBC_i	$x_B < x_i < x_C$, means bus i is between point B and C.
XC_i	$x_i = x_C$, (or $|x_i - x_C| < \varepsilon$), means bus i is arrives at stop C.
XCD_i	$x_C < x_i < x_D$, means bus i is between stop A and point B.
XD_i	$x_i = x_D$, (or $|x_i - x_D| < \varepsilon$),means bus i is arrives at stop D.
Forward$_i$	$dir_i = 1$, means bus i is if moving from left to right.
Stop$_i$	Bus i is decelerating/braking to stop.
Go$_i$	Bus i is accelerating to/running at its desired speed.

true and Go$_i$ to be false are listed in this table; in all the other conditions, Stop$_i$ will be false and Go$_i$ will be true.

The whole bus-road system has been simulated in MATLAB. Some experiment results are shown in Figure 3.17.

In the following experiment, one bus starts from $x = 0.2$ mile (on AB), running towards D, and another one starts from $x = 13.0$ miles (on CD), running toward A. The two buses then just move back and forth between stops A and D.

Several parameters in the experiment are chosen as follows:

- $M = 20$ ton.
- $V_0 = 80$ mph (for bus 1) and 70 mph (for bus 2).

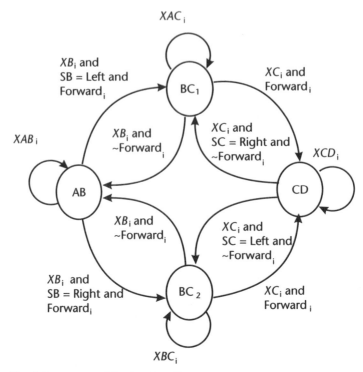

Figure 3.15   The finite state machine for state P_i.

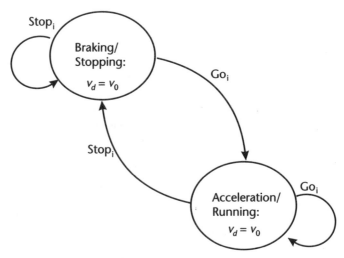

Figure 3.16   The finite state machine for state V_i.

- $K_1 = -10,000$ ton/hr.
- The sampling frequency is 1 Hz.
- The distances between AB, BC, and CD are all 5 miles.
- The range of the sensor is 5.0 miles (in order to guarantee no collisions).

Table 3.10   The Truth Table for Events Stop and Go for the Train

P_i	dir_I	P_o	dir_o	$Stop_i$	Go_i
BC_1	1	CD	−1	True	False
BC_1	−1	AB	1	True	False
BC_2	1	CD	−1	True	False
BC_2	−1	AB	1	True	False

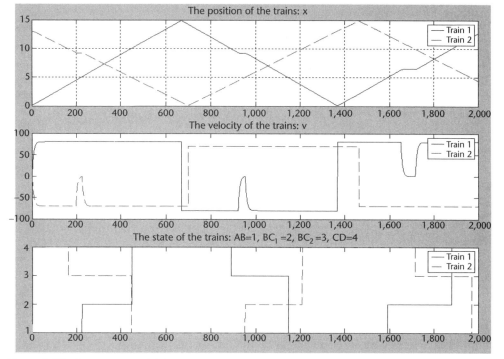

Figure 3.17   Bus system variables.

We can see, as shown in Figure 3.17, that with our control system the two buses can run on the route without any collisions.

3.3   State Machines for Different Challenge Events

3.3.1   Macrostates: Highway, City, and Off-Road Driving

It has to be pointed out that the two examples provided expose the full state machines that control all the micro-operations. Fully automated driving will have numerous operations such as these, and will obviously be controlled by a higher-level state machine. We shall call the states of such a machine the *macrostates*. Each macrostate will represent many substates and the transitions among them.

Studies on the development of AHS usually advocate an ingress-egress pair during which the car will follow the assigned lane on the highway. Early testing and demonstration implementations, for example in Demo '97, assumed that the cars would be following specialized technological aids indicating the precise location of the car with respect to the lane. Figure 3.18 shows one technology that

Figure 3.18 Two autonomous cars developed by OSU in Demo '97 following a radar-reflecting stripe and undertaking a pass.

was advocated for location information with respect to the lane, a radar-reflecting stripe [6] that would indicate the distance from the center of roadway and the relative orientation of the car. We will mention other possible technologies later in this book. It has to be pointed out that precision GPS and maps were not commonly available at that time. Today, it is assumed that precision maps would be available to the level of identifying individual lanes, and GPS reception would provide precise location information in real time.

Demo '97, one of the early full-scale AHS demonstrations, was held on a 7.5-mile segment of highway I-15 in San Diego. This segment was a segregated two-lane highway normally used for rush-hour high-occupancy vehicle traffic. Traffic flowed in the same direction in both lanes, and there were no intermediate entry and exit points. The curvature of the highway lanes was benign and suited for high speed (70 mph) driving, and other traffic was minimal to nonexistent. A general AHS would presumably have merge and exit lanes, but the single entry-exit aspect of Demo '97 made it a single activity: *drive down the lane and possibly handle simple interactions with other vehicles.* We shall subsequently call this behavior a *metastate*. Dealing with interchanges produced by entry and exit lanes would require other metastates.

The DARPA Grand Challenges of 2004 and 2005 were both off-road races. As such, the only behavior and thus the only metastate required would be path following with obstacle avoidance from point A to point B. However, since there is no path or lane that can be discerned from a roadway, the only method of navigation is to rely on GPS- and INS- based vehicle localization [7, 8] and a series of predefined waypoints. Obstacle avoidance would be needed, as in an AHS, although in the less structured off-road scenario greater freedom of movement and deviations from the defined path are allowed. The Grand Challenge race rules made sure that there were no moving obstacles and different vehicles would not encounter each other in motion. General off-road driving would of course not have this constraint.

Finally, fully autonomous urban driving would introduce a significant number of metastates—situations where different behavior and different classes of decision need to be made. The DARPA Urban Challenge, although quite complex, did have fairly low speed limits, careful drivers, and no traffic lights. Visual lane markings were unreliable, and thus true to life, and the terrain was fairly flat, although some areas were unpaved, generating an unusual amount of dust and creating problems for some sensors. See Figure 3.19 for different types of vehicles used in the Grand and Urban Challenges.

We will make the claim that the basic problem definition will include a series of waypoints and a concept of lanes. Although lanes are obvious in highway systems and urban routes, it is reasonable to assume that off-road environments also present a set of constraints that indicate the drivability of different areas and thus provide the possibility of defining lanes.

We indicated in Chapter 1 that we would assume a basic setup with waypoints and lane boundaries. The off-road situation where the feasible path is understood to be a lane provides the simplest illustration of metastates (see Figure 3.20). A highway configuration would have multiple lanes all headed in the same direction, with standard lanes of equal width.

On the other hand, both AHS and urban automated driving scenarios need concepts/tasks related to changing lanes. One possible metastate configuration is shown in Figure 3.21.

The urban environment requires a much more complex state machine to represent the situations and control transitions [9]. Figure 3.22 illustrates the metastates used in OSU-ACT. It has to be pointed out that Figure 3.22 hides many more substates underneath, as compared to the metastates of Figure 3.20 (see [10]).

3.3.2   The Demo '97 State Machine

3.3.2.1   The World and Personalities

Consider first a "world" comprised of a single lane. The decision making here is quite simple. Any car, with a specified speed and headway will just need to check for

Figure 3.19 Autonomous vehicles developed by OSU. (a) ACT at UC '07; (b) TerraMax at GC '04; and (c) ION at GC '05.

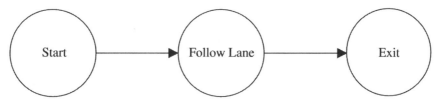

Figure 3.20 A three-metastate model of transporting from start to end (exit).

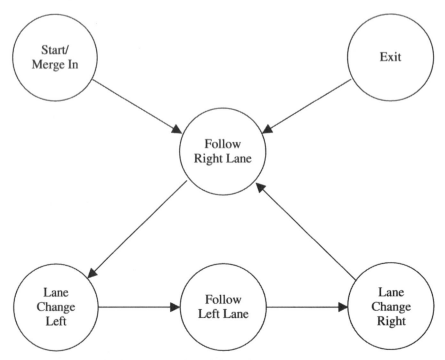

Figure 3.21   A two-lane highway example with lane change.

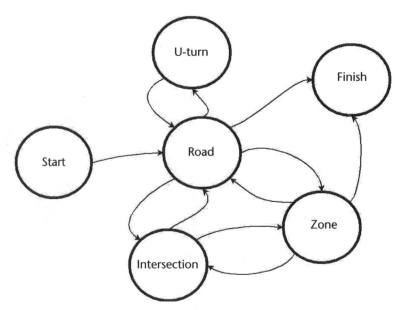

Figure 3.22   Metastates for the 2007 DARPA Urban Challenge situation.

a car ahead. Once following the car ahead, we can consider a number of different reactions if it changes speed.

We shall now introduce the concept of "personality" to represent a set of behavior patterns. This set will affect the decisions of the car. At this initial stage we shall consider four classes:

1. Grandparent;
2. Teenager;
3. Jack;
4. The homicidal driver.

With the single-lane world constraint, the first three personalities do not provide much variation. The teenager may be more aggressive in its pursuit of the car ahead. Other combinations can easily be analyzed. With many cars on the single lane, it is obvious that eventually all traffic would accumulate behind the grandparent unless there is a homicidal driver. Assuming such a driver could reverse direction, this creates many unmanageable situations.

When we expand our world to two lanes and allow passing, even with the first three personalities, a number of interesting situations arise. We will make a set of assumptions about the personalities.

- Grandma drives slow. Never passes.
- Teenager follows slow car ahead, but has two specific characteristics that distinguish him/her from Jack: A short "boredom time," and a low threshold risk analysis. (Teenager will change lanes with a narrow merge distance.)
- Jack can show variations in his driving pattern, staying within his personality.

The last item above allows us to consider manual drivers together with automated vehicles. Indeed, we have assumed a manual grandparent, an automated teenager, and Jack, during our runs in the 1997 Technology Demonstration. It was assumed that the grandparent could speed up, slow down, and even stop. He or she would not do a sharp stop, change lanes, or speed more than a limit (less than the teenager and Jack). The analysis of the scenario and the hybrid system model is provided next.

3.3.2.2   OSU Demo Scenario

The scenario involves three cars, two of which are fully automated and the other is manually driven. The demo took place on a two-lane, unidirectional, 7.6-mile-long road segment in Southern California. The vehicles in the scenario are referred as the "Grandma" (manually driven car), the "Teenager" (automated car 1), and "Jack" (automated car 2). The OSU Demo '97 scenario is described in detail in [2, 6]. Additionally, we state that, as far as the controllers, sensors, and actuators are concerned, all automated cars (the two in the demo and the spare) are identical. Hence, their roles, in the demo, are interchangeable upon modifying a couple of parameters which describe the "personality" of the vehicle.

Each automated vehicle is capable of performing headway (slow down, speed up, cruise, stop) and lateral control (lane keeping, lane change).

3.3.2.3   Discrete Event System Model

The discrete event system portion of the hybrid system is modeled as a finite state machine, whose state transitions are manipulated by some events that occur in the

CSS. These events are processed in the CSS (with filters and observers) and are passed to the DES side through an interface.

Let the set of low-level events and requests be denoted by the set E (see Table 3.11). There is one discrete state variable X. See Table 3.12 for a list of those states.

The discrete system state transition function shows the low-level events passed on to the DES side. Lateral and longitudinal states are listed in Tables 3.13 and 3.14.

3.3.2.4   Interface Layer

Recall that the objective of the controller design is to develop a DES supervisor to regulate the system dynamics to the desired trajectories. The interface layer has

Table 3.11   The List of Low-Level Events and Requests

e_1^1	Manual request
e_1^2	Try auto request
e_2^1	Start request
e_2^2	Stop request
e_3^1	Left lane change request
e_3^2	Right lane change request
e_4^1	Auto zero request
e_5^1	Target acquired ahead
e_5^2	No target acquired ahead
e_6^1	Ready for lane change
e_6^2	Vehicle following mode

Table 3.12   Top-Level Discrete States

e_1^1	Manual request
e_1^2	Try auto request
e_2^1	Start request
e_2^2	Stop request
e_3^1	Left lane change request
e_3^2	Right lane change request
e_4^1	Auto zero request
e_5^1	Target acquired ahead
e_5^2	No target acquired ahead
e_6^1	Ready for lane change
e_6^2	Vehicle following mode

Table 3.13   Lateral DES States

Lane keep state

Lane change start state

Lane change complete state

Fail to manual state

Table 3.14   Longitudinal DES States

Starting state

Stop state

Preslowing state

Slowing state

Preacceleration state

Acceleration state

Precruising state

Cruising state

Prefollowing state

Following state

the major task of building the bridge between the continuous state system and the discrete state system within the hybrid model. Based on the suggested control law, certain continuous input is selected from a finite set of all possible inputs to fully automate the motions of the considered vehicle. The overall layout of the OSU system is given in Figure 3.23 [9].

We now consider the finite state machine that will control the scenario. The states are defined in Table 3.15 and the state machine is also provided in Figure 3.24.

The Demo '97 scenario provides an interesting example of different behavior in hybrid systems where the threshold values in the interface are selected differently. The so-called "boredom factor" adjusting the time a car will continue following, before starting a lane change so as to pass, are set different in the two cars. As intended, this results in the teenager passing, but Jack not passing the grandparent.

Our Demo '97 scenario did not include the possibility of another car approaching from the left lane, a standard situation that may lead to unsafe situations in the real world. Normally, the lane-changing car would check for oncoming vehicles in the left lane.

3.3.3   Grand Challenge 2 State Machine

Figure 3.25 represents the FSM we implemented in ION [7]. In this FSM, we centered the state *path-point keeping* and *obstacle avoidance* to suit the task of GC05, in which the path-planning algorithm is activated. In the *tunnel mode* state, we tried to put the vehicle in the center of the tunnel by measuring the distance to both sides from ultrasonic transducers mounted on both sides of the vehicle. When something goes wrong (for example, a sensor failure happened twice in our GC05 race), the FSM gets in the *alarm* state to deal with the malfunction by resetting the sensor. The

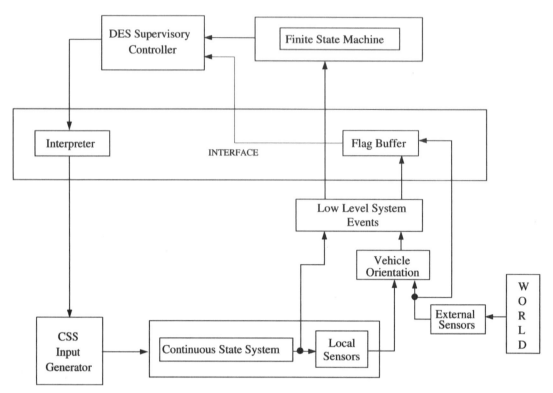

Figure 3.23 OSU Demo '97 system.

Table 3.15	List of States
MAN	Manual mode
RLC	Right lane change
ACCinR	Accelerate in the right lane
CRUinR	Cruise in the right lane
DECinR	Decelerate in the right lane
FOLinR	Follow in the right lane
LLC	Left lane change
ACCinL	Accelerate in the left lane
CRUinL	Cruise in the left lane
DECinL	Decelerate in the left lane
FOLinL	Follow in the left lane

road following, rollback, and *robotic operations* states work when the conditions are met.

To prevent the FSM from being stuck in some states other than path-point keeping and obstacle avoidance forever, a watchdog is introduced. When the vehicle keeps still or the FSM stays in some unwanted state for a certain period of time and no promising progress is expected, the FSM and some modules reset.

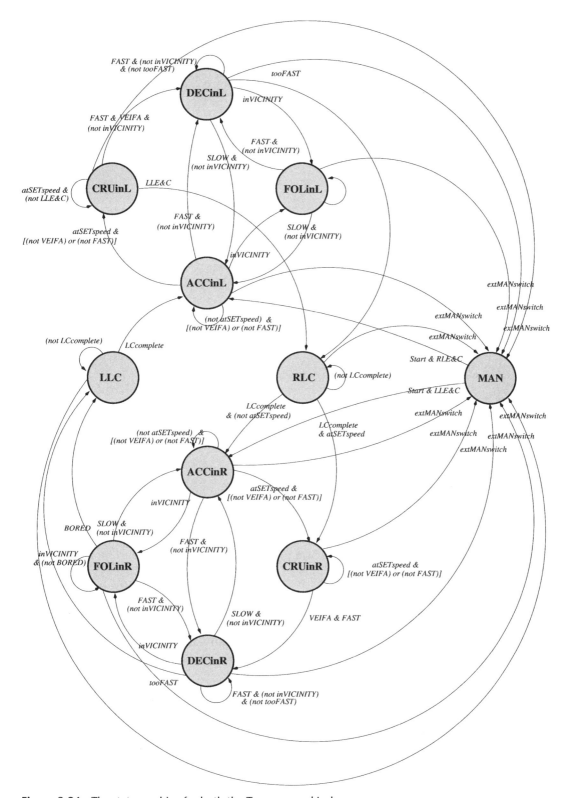

Figure 3.24 The state machine for both the Teenager and Jack.

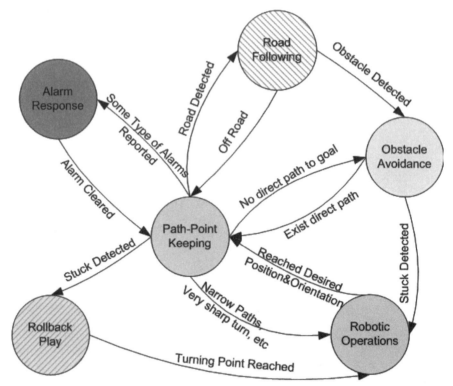

Figure 3.25 The FSM for ION participating in the second Grand Challenge.

3.3.4 The Urban Challenge State Machine

The DARPA Urban Challenge provided a much richer environment with many possible situations, as compared to the DARPA Grand Challenges. We therefore structured the FSM into a two-level hierarchy in which the higher-level states were labeled metastates (Figure 3.22).

As examples, we provide a series of metastates that were originally developed for the Urban Challenge. (These are provided as examples and are not necessarily those used in the actual event.)

3.3.4.1 On a One-Lane Road

In this exposition we distinguish between one-lane and two-lane roads, the basic difference being in the lane change to pass operation (see Figure 3.26).

List of events:

E1: Entering a one-lane road.

E2: Slow traffic ahead, distance < 30m.

E3: Slow traffic ahead stopped.

E4: Slow traffic ahead, distance < 8m; ~ E4: distance > 8m.

E5: Traffic cleared.

E6: Close to stop line, < 10m.

E7: Reached the stop line, < 1m.

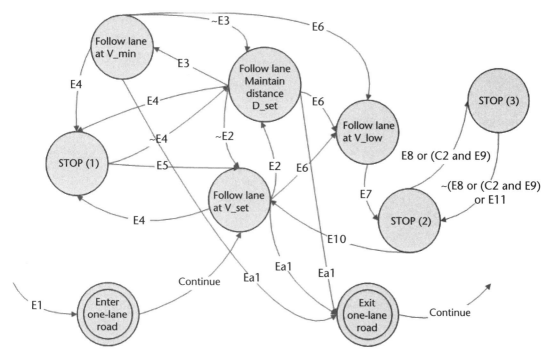

Figure 3.26   The one-lane road metastate.

E8: Exists precedent stopped vehicle.

E9: Traffic in the crosslane is within 10 seconds.

E10: Intersection cleared.

E11: Precedent vehicles don't move > 12 seconds + rand() after E10.

Ea1: Reached the Exit_Point.

C1: There is a stop sign at the exit point.

C2: The crosslane doesn't have stop sign.

3.3.4.2   At a T-Junction Exit

There are four types of T-junction exit operations, as shown in Figure 3.27, leading to the states shown in Figure 3.28.

E18: Reached a T-junction exit point.

E19: Goal lane empty > 8m from previous vehicle.

E20: No traffic in lane A within 10 seconds.

E21: No traffic in lane B within 10 seconds.

E22: Reached goal lane.

C4: STOP sign at position A.

C5: STOP sign at position B.

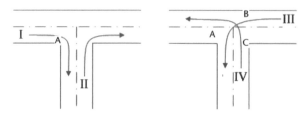

Figure 3.27 T-junction operations.

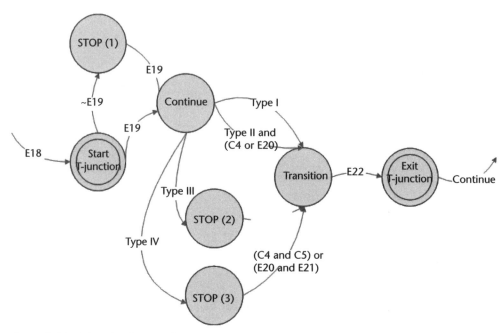

Figure 3.28 T-junction intersection states.

3.3.4.3 The U-Turn

We detect a U-turn if (a) exit_point and enter_point has the same segment_id and/ or (b) the drive directions are opposite to each other. The scenario is as shown in Figure 3.29 and the states are in Figure 3.30.

The events in the U-Turn state machine are listed here:

E25: Reached a U-turn exit point.

E26: Vehicle in lane B.

E27: θ_diff is small.

E28: Vehicle reached center of lane B.

E29: θ_diff > θ_min.

E30: Stopped.

Ea2: Detected object in the near back.

θ_diff: the heading difference between vehicle and lane B.

θ_min: Parameter, acceptable heading difference.

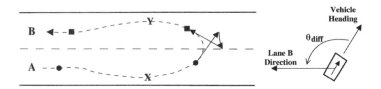

Figure 3.29 The U-turn.

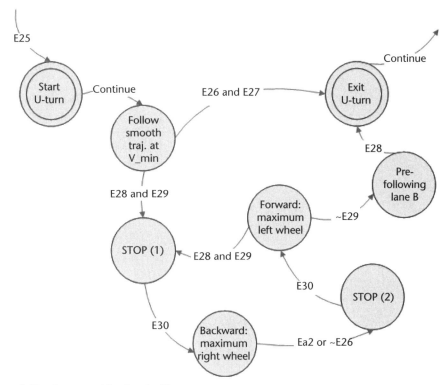

Figure 3.30 State machine for the U-turn.

References

[1]   Özgüner, Ü., and K. Redmill, "Sensing, Control, and System Integration for Autonomous Vehicles: A Series of Challenges," *SICE Journal of Control, Measurement, and System Integration*, Vol. 1, No. 2, March 2008, pp. 129–136.

[2]   Redmill, K., and Ü. Özgüner, "The Ohio State University Automated Highway System Demonstration Vehicle," *SAE Transactions 1997: Journal of Passenger Cars*, sp-1332, Society of Automotive Engrs., 1999.

[3]   Özgüner, Ü., C. Stiller, and K. Redmill, "Systems for Safety and Autonomous Behavior in Cars: The DARPA Grand Challenge Experience," *Proceedings of the IEEE*, Vol. 95, No. 2, February 2007, pp. 397–412.

[4]   Lygeros, J., D. N. Godbole, and S. Sastry, "A Verified Hybrid Controller for Automated Vehicles," *Proceedings of the 1996 Conference on Control and Decision*, Kobe, Japan, 1996, pp. 2289–2294.

[5]   Passino, K., and Ü. Özgüner, "Modeling and Analysis of Hybrid Systems: Examples," *Proc. 1991 IFAC Symp. on DIS & 1991 IEEE International Symp. on Intelligent Control*, Arlington, VA, Aug. 1991.

[6]   Özgüner, Ü., et al., "The OSU Demo '97 Vehicle," *IEEE Conference on Intelligent Transportation Systems*, Boston, MA, November 1997.

[7]   Chen, Q., Ü. Özgüner, and K. Redmill, "Ohio State University at the 2004 DARPA Grand Challenge: Developing a Completely Autonomous Vehicle," *IEEE Intelligent Systems*, Vol. 19, No. 5, September-October 2004, pp. 8–11.

[8]   Chen, Q., and Ü. Özgüner, "Intelligent Off-Road Navigation Algorithms and Strategies of Team Desert Buckeyes in the DARPA Grand Challenge '05," *Journal of Field Robotics*, Vol. 23, No. 9, September 2006, pp. 729–743.

[9]   Özgüner, Ü., C. Hatipoglu, and K. Redmill, "Autonomy in a Restricted World," *Proc. of I. IEEE ITS Conf.*, Boston, MA, November 9–12, 1997, p. 283.

[10]   Kurt, A., and Ü. Özgüner, "Hybrid State System Development for Autonomous Vehicle Control in Urban Scenarios," in C. Myung Jin and M. Pradeep, (eds.), *Proceedings of the 17th World Congress The International Federation of Automatic Control*, Seoul, Korea, 2008, pp. 9540–9545.

CHAPTER 4

Sensors, Estimation, and Sensor Fusion

Sensors are applied in all levels of vehicle control and autonomy, ranging from engine control, ABS braking and stability enhancement systems, passive driver assistance systems such as navigation, infotainment, backup hazard warning, and lane change assistance, active safety systems such as lane maintenance and crash avoidance, and, of course, full vehicle automation.

In broad terms, sensors can be grouped according to the function they provide. Internal vehicle state sensors provide information about the current operation and state of the vehicle, including lower-level functions such as engine operations and higher-level states such as vehicle motion and position. External environment sensors provide information about the world outside the vehicle, potentially including road and lane information, the location and motion of other vehicles, and stationary physical objects in the world. Finally, driver state and intention sensors provide information about the state or intentions of the driver. These sensors can include seat occupancy and passenger weight (pressure or infrared sensors), audio sensors, internal cameras, eye trackers, breath alcohol sensors, and haptic transducers.

In this chapter, we will review the general characteristics of sensors and sensor performance. Then we will look at the individual sensors and technologies that are generally applied for vehicle control and automation. Since it is often both advantageous and necessary to combine the information from multiple sensors to provide a full and error-free understanding of the current state of the vehicle and the world, we end this chapter with a description of estimation and sensor fusion approaches.

Conceptually, any device or technology that provides information can be considered, and treated, as a sensor. In vehicle automation applications, common examples of this include map databases and wireless vehicle-to-vehicle and vehicle-to-infrastructure communications, which are discussed in Chapters 6 and 7, and other cooperative infrastructure technologies including visual signs, tags, or markers and radar reflective surfaces, which are discussed in this chapter.

4.1 Sensor Characteristics

Sensors are fundamentally transducers in that they convert one physical property or state to another. There are several general characteristics that are important in describing and understanding the behavior of sensors and sensor technologies. While we will not cover these in detail, it is worth bearing in mind how these factors relate to the selection, interpretation, and fusion of sensors individually and when used as a sensor suite.

- *Accuracy:* The error between the true value and its measurement, which may include noise levels and external interference rejection parameters;
- *Resolution:* The minimum difference between two measurements (often much less than the actual accuracy of the sensor);
- *Sensitivity:* The smallest value that can be detected or measured;
- *Dynamic range:* The minimum and maximum values that can be (accurately) detected;
- *Perspective:* Quantities such as the sensor range or its field of view;
- *Active versus passive:* Whether the sensor emits energy or radiation that illuminates the environment or relies on ambient conditions;
- *Timescale:* Quantities such as the update rate of the sensor output and the frequency bandwidth of the measurement output over time;
- *Output or interface technology:* For example, analog voltage or current, digital outputs, and serial or network data streams.

4.2 Vehicle Internal State Sensing

Autonomous vehicles use all the standard sensors available in a car for self-sensing. Thus speed sensing is available, and variables that are measured in the engine and powertrain or are related to the brakes can be accessed. Furthermore, sensors are needed to measure steering wheel angle and gear shift, but these are fairly easy to design or develop if not already present on the vehicle.

4.2.1 OEM Vehicle Sensors

Modern vehicles have sophisticated electronic control systems that require a number of sensors and measurements. The outputs of these sensors may appear on the vehicle's internal communication bus, or they be electronically tapped for use. Commonly available measurements include:

- Wheel speed, usually measured by a Hall effect sensor, which produce a digital signal whose frequency is proportional to speed;
- Vehicle dynamic state, possibly including yaw rate and lateral and longitudinal acceleration;
- Driver inputs, for example, steering wheel position, throttle and brake pedal positions, turn signals, headlights, windshield wipers, and so forth;
- Transmission gear and differential state;
- Brake pressure, either at the master cylinder or for each wheel, usually measured by a diaphragm or silicon piezoelectric sensor;
- Engine and exhaust variables, for example, coolant temperature, O_2 and NOX levels, RPM, and spark plug firing timing.

Since these sensors are designed by the vehicle OEM to serve a specific purpose in the vehicle (for example, ABS, stability enhancement, or powertrain control), and given the significant price pressure in the automotive market, these sensors tend to be only as good as is required for their designed application. They are not always of sufficient quality for vehicle automation.

4.2.2   Global Positioning System (GPS)

A Global Positioning System (GPS) is a key component of a present day intelligent vehicle. GPS can be used to find absolute position and velocity. This is very useful for an autonomous vehicle that has access to a precise map, as it can understand where it is with respect to its destination and with respect to the road network or, for an off-road vehicle, topographic features and obstacles. This information is needed, for example, to compute optimal routes or driving directions. On- or off-road, when combined with a vehicle-to-vehicle wireless communication system, it can also provide relative position and relative speed information.

The information available from a GPS receiver, such as that shown in Figure 4.1, is:

- Absolute position in a geodetic coordinate system, for example, latitude-longitude-altitude, X-Y-Z Earth centered Earth fixed, or UTM;
- Velocity and course over ground information (horizontal speed and orientation relative to true north);
- Precise time and synchronized pulse per second;
- Raw information that can be used for precise postprocessing applications.

Various level of accuracy in GPS measurements can be obtained, depending on the types and number of GPS signals received and analyzed, the algorithms, and the availability of externally supplied correction data (including differential correction and real-time kinematic correction data). Satellite visibility and geometric configuration is also a significant factor in GPS measurement accuracy, especially in areas with tree foliage coverage or buildings that can occlude signals from certain satellites or create multipath reflections.

Generally speaking, in areas with an unobstructed view of the entire sky and barring a particularly bad orbital configuration of visible satellites, a standard

Figure 4.1   A GPS receiver and antenna. (Figure courtesy of Novatel.)

inexpensive or embedded GPS receiver can achieve position accuracies on the order of 5–15 meters. This is sufficient for providing navigation and routing instructions for a human driver, but is insufficient for resolving which lane a vehicle is currently occupying.

To achieve the next level of GPS position accuracy, correction signals are available from free and publicly available services, such as the Nationwide Differential GPS (NDGPS) [1] service broadcasting in the long-wave band and satellite-based augmentation systems (SBAS) such as the Wide Area Augmentation System (WAAS) provided by the U.S. Federal Aviation Administration or the European Geostationary Navigation Overlay Service (EGNOS). An appropriately capable GPS receiver, using one of these basic differential correction data services broadcast over a large area of the planet, can achieve position accuracies on the order of 1–2 meters. This is sufficient for some safety applications, and can sometimes resolve lane identity, but is insufficient for autonomous vehicle operations.

Locally computed differential corrections or commercial SBAS systems, for example, Omnistar VBS and Firestar, can achieve submeter accuracies, which significantly improves the performance of data collection systems and safety systems, and in some cases is sufficient for longitudinal vehicle automation.

The more sophisticated commercial correction services such as Omnistar HP can achieve position accuracies of 10 cm or better. High-precision locally computed real-time kinematic correction systems can produce position accuracies of 1–2 cm. Measurements at this level of accuracy are sufficient for full vehicle automation. Of course, both the correction data and the GPS receiver hardware needed to achieve these levels of accuracy are quite expensive.

The time information from the GPS is particularly useful in intervehicle communication since it allows the precise synchronization of clocks across multiple vehicles and infrastructure systems.

4.2.2.1 Fundamentals of GPS

The GPS constellation, as illustrated in Figure 4.2, consists of approximately 24 satellites arranged in 6 orbital planes at an altitude of 12,500 miles (20,200 km). Each satellite completes 2 orbits each sidereal day. Each satellite broadcasts at least the following signals and information:

- L1 frequency: 1,575.42 MHz (19 cm) at 25.6W;
- L2 frequency: 1,227.60 MHz (24 cm);
- Course acquisition (C/A) code (on L1):
  - 1.023 megabits/sec PRN (pseudorandom noise);
  - 1,023 bits long (repeats every 1 ms, 300 km);
- Precise (P,Y) code (on L1, L2, 90° out of phase):
  - 10 MHz PRN, encrypted;
  - Approximately 6 trillion bits long (repeats every 7 days);
- Navigation message:
  - 50-Hz data;
  - 12-minute message time;

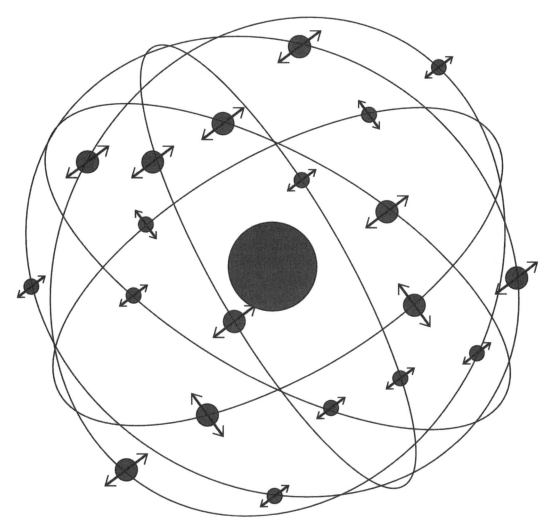

Figure 4.2   GPS satellite constellation [2].

- Subframe data:
  - UTC time and clock corrections;
  - Almanac;
  - Precise ephemeris data (30 seconds);
  - Ionospheric propagation data.

Newer GPS satellites support signals transmitted on additional frequencies, for example, on the L5 (1,176.45 MHz) band.

GPS technology is, fundamentally, based on triangulation and the fact that, in terms of RF signal propagation, distance travelled is directly related to elapsed time (in a vacuum 3.3 ns is approximately equivalent to 1 meter). Consider that any object that is known to be a distance D from a given point must lie on the surface of a sphere of radius D centered on that point. Two noncoincident points with known distances from the object define two spheres, and the object must lie on the intersection of those spheres, which is a circle except in special degenerate cases.

Three points generally reduce the possible location of the object to two points, one of which is, in the case of the GPS system, unfeasibly far from the surface of the Earth. This is illustrated in Figure 4.3.

Thus, in order to compute the position of an object we need its distance to at least three known points. The orbital location of a GPS satellite can be computed fairly precisely using the almanac and ephemeris data transmitted by the satellites themselves coupled with a mathematical model of orbital dynamics. Indeed one of the primary functions of the GPS ground control system is to track the satellites and regularly update the almanac and ephemeris information. What remains is to determine the distance, or range, from each of at least three satellites to the antenna of the GPS receiver.

As mentioned earlier, the distance travelled is a direct function of the elapsed time between the transmission of a signal and its reception. GPS satellites carry atomic clocks of very high precision, and the GPS ground control system ensures that the clocks of all satellites are synchronized to within a few nanoseconds. The data transmitted by each GPS satellite also contains information about when transmissions occur. All GPS satellites transmit on the same frequencies, so each satellite transmits its data stream by modulating its own unique binary sequence pseudo-random noise (PRN) code. On the L1 channel, the PRN codes are 1,023 bits long and repeat every millisecond. Since the GPS receiver knows the PRN sequence of each satellite, it can produce its own copy of that sequence synchronized to its internal clock and use correlation to determine the time offset of the received and internally generated signals for each visible satellite. This is illustrated in Figure 4.4.

However, GPS receivers are typically low-cost devices using nothing more than a crystal-controlled clock with considerable drift over time relative to the GPS satellite clocks. Thus, it is necessary to align the receiver clock with the satellite clocks. This introduces a fourth unknown into the problem, and therefore a full GPS position fix actually requires four satellites. Since the clock bias (the difference between the receiver's clock and the satellite clock) is unknown, the distance or range measured by the PRN correlation process is called a pseudorange.

Expressed mathematically in Earth centered Earth fixed coordinates, the measured pseudorange for each visible satellite is

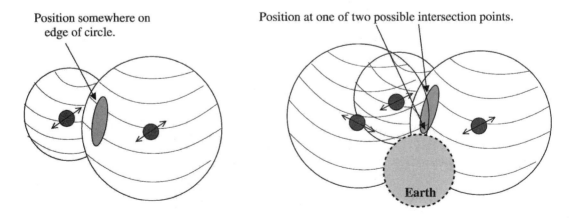

Figure 4.3 Position from intersection of multiple spheres (triangulation).

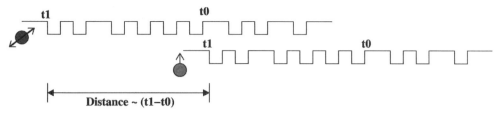

Figure 4.4   Pseudorange from temporal correlation.

$$P_i = \sqrt{(X_i - x)^2 + (Y_i - y)^2 + (Z_i - z)^2} + ct + v_i$$

where

(X_i, Y_i, Z_i) is the ECEF position of the satellite;

(x, y, z) is the ECEF position of the observer;

ct is the clock bias (c is speed of light);

v_i is an unknown offset to due errors and noise;

$i = 1...n$ where n is the number of visible satellites.

With $n >= 4$ satellites, this forms a system of possibly over-determined nonlinear equations that can be solved for the four unknowns using a variety of numerical techniques to obtain the (x, y, z) position coordinates and the clock bias ct at the antenna of the GPS receiver. As an example, one possible solution technique involves linearizing the pseudorange equations, using an estimate of (x, y, z, t) and a Taylor's series approximation, and solving using an SVD/pseudoinverse approach or a Newton-Raphson iterative approach.

An overall block diagram of a basic GPS receiver is shown in Figure 4.5. The filtered, amplified, and downconverted signals received from the GPS satellites are correlated with locally generated PRN codes to generate pseudorange values for each satellite in view. The data encoded in the satellite transmissions is then combined with the pseudorange values to generate a navigation solution for the current position and velocity. Of course, more sophisticated GPS receivers may contain additional functional blocks.

4.2.2.2   GPS Error Sources

Considering the above, it is clear that errors in the final position solution can arise from errors in any of the fundamental measurements, including satellite position, clocks and timekeeping, and the pseudorange distance measurements. A basic error budget (1-sigma values) for a standard GPS receiver would be:

- *Broadcast ephemeris (satellite position and orbit data):* 2.5 meters;
- *Satellite clocks:* 1.5 meters;
- *Ionosphere delays:* 5.0 meters;
- *Troposphere delays:* 0.7 meters;

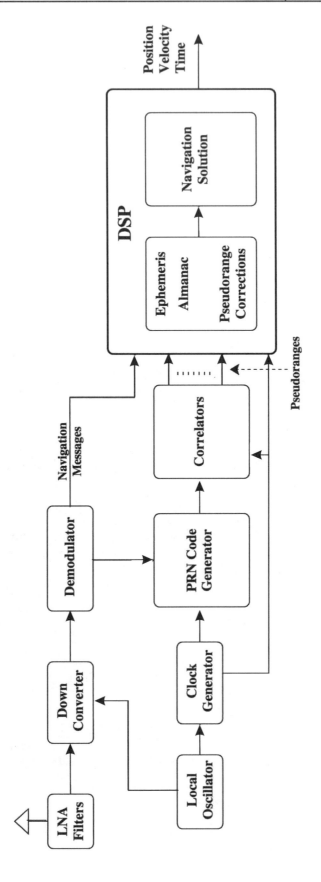

Figure 4.5 Generalized block diagram of GPS receiver.

- *Receiver noise and numerical computations:* 0.7 meter;
- *Multipath and diffraction:* 1.5 meters.

It is worth mentioning that many of these errors can be considered as slowly time varying (autocorrelated) signals or almost as bias terms, so is it common to see GPS position errors slowly wander within a region surrounding the true position.

Given the actual position and the cloud of GPS measurements from the receiver, there are two different accuracy measures that are used in the industry:

- *Circular error probability (CEP):* The radius of the circle that is expected to contain 50% of the measurements.
- *Twice the distance root mean square (2dRMS):* The radius of the circle that is expected to contain 95% of the measurements.

GPS manufacturers tend to report the expected performance of a given receiver using one of these two methods.

Another source of uncertainty in the performance of a GPS receiver is related to the configuration of satellites that a receiver can see at a given time. If the receiver has a clear view of the entire sky and the satellites are widely distributed throughout the sky, then the region of possible position solutions is geometrically compact (Figure 4.6).

On the other hand, if the sky is partially occluded by buildings, trees, or other structures, or if the satellites happen to be in a cluster at a given time, then the regions of possible position solutions may be irregularly shaped and large (Figure 4.7). This phenomenon is called geometric dilution of precision, and most GPS receivers will provide unitless measurements of its severity at any given time.

It is important to note that GPS sensors require external information provided by radio communication with satellites. Some or all satellites can be obscured by buildings or when in tunnels, and this may cause degradation in performance or a complete loss of sensor function at any time. It is also possible to jam GPS signals, which are quite weak. It is possible to continue to provide position estimates by fusing GPS information with other sensors, but the accuracy of these position

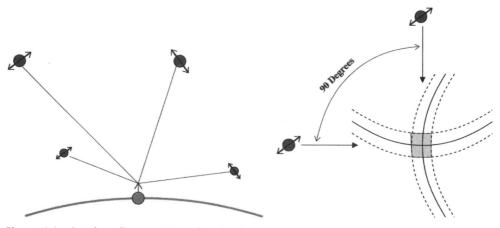

Figure 4.6   Good satellite geometry—low DOP.

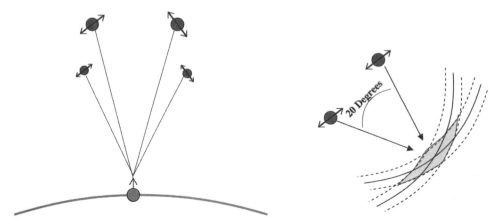

Figure 4.7   Poor satellite geometry—high DOP.

estimates will degrade over relatively short time spans. It may also be possible to continue autonomous vehicle operation using SLAM techniques or only local information, for example, lane marker sensing in the case of driving down a limited access highway.

This continues to be an issue for autonomous vehicle implementations as well as an area of active research, since a loss of accurate position and orientation information adversely effects most other vehicle sensing and control systems.

4.2.2.3   Improvements to Standard GPS Performance

A number of the significant measurement errors that contribute to GPS position errors are well correlated over a local geographic area. That is to say, if two GPS receivers are relatively close together, they will both observe the same set of satellites and the propagation paths that the RF signals follow from the satellites to the receivers will pass through the same atmospheric conditions. This fact is the basis for various forms of differential correction.

A high-quality GPS receiver installed at a precisely known position, often called a base station, can, instead of computing its own position, use the pseudorange equations and error models to compute the current values of the errors present in the signals from each satellite. This information can be relayed, as shown in Figure 4.8, usually through a wireless data link, to a remote GPS receiver, often called a rover, which can use this information to remove the known errors from its own measurements and position computation. The quality of this differential correction depends on how close the rover is to the base station. Generally, significant improvements can be seen for baseline lengths up to 150 km.

One can purchase and install a GPS differential base station for ones own use. There are a number of national governments that provide freely available differential GSP corrections, usually transmitted as long wave (275–325 kHz) radio signals, however these services tend to be clustered around maritime features and may not provide universal land coverage. In the United States, the Federal Highway Administration has been expanded the older U.S. Coast Guard service to provide a more uniform national coverage under the NDGPS program [1]. Satellite-based augmentation systems (SBAS) and wide-area differential corrections (WADGPS)

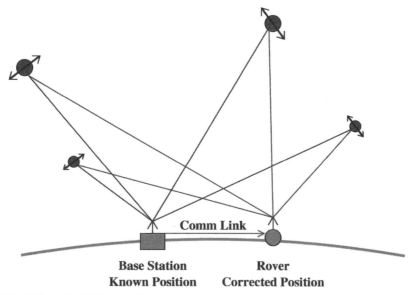

Figure 4.8   Differential GPS corrections.

providers use many base stations covering, for example, an entire continent to model GPS observation errors and provide the data to GPS receivers often as satellite broadcasts. Examples include the U.S. FAA's WAAS system and the European EGNOS system, which can generally increase position accuracy to within 1–2 meters of truth, as well as commercial providers such as Omnistar and Racal, which offer varying levels of services down to the 10-cm range.

Another approach to improving GPS position accuracies involves measuring the actual carrier phase of the received GPS satellite transmissions, as shown in Figure 4.9, to provide higher accuracy range measurements. Very small changes in carrier phase can be tracked over time (the L1 carrier wavelength is 19 cm). This is most effective when the GPS receiver is capable of receiving both L1 and L2 transmissions, since the different propagation characteristics of the two frequencies can be used to identify and quantify errors.

Some high-end GPS receivers can also measure Doppler shifts in order to directly compute the relative velocities of each satellite. Multiple GPS receivers can also be combined with specialized algorithms to measure absolute yaw angle.

Finally, it is possible to store the raw code and carrier phase measurements (and possibly the Doppler velocity measurements) logged by a remote GPS receiver, and to obtain precise satellite ephemeris data available after the fact (usually within 6–24 hours), in order to compute highly accurate position and velocity

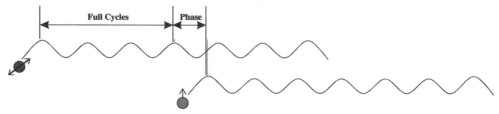

Figure 4.9   Carrier phase measurements.

measurements, in some cases with subcentimeter accuracy. This process is known as *postprocessed GPS*.

It should be noted that commercial GPS receivers, especially high-precision GPS receivers, employ many sophisticated algorithms and approaches, including carrier phase and dual frequency measurements, multipath identification and mitigation, Kalman filters, and other proprietary technologies to improve GPS position accuracy. These, along with more complicated correction technologies such as real-time kinematic (RTK) GPS, are beyond the scope of this text.

4.2.3 Inertial Measurements

When a vehicle is in motion, it may experience linear and rotational motions along each axis: lateral, longitudinal, and vertical. Inertial measurements, which can be made internally without reference to an external point, include linear acceleration and angular rates and accelerations. This information can be used to augment or improve external measurements such as GPS coordinates.

Accelerations along these axes can be measured using an accelerometer. Commonly available accelerometers measure the force generated by the accelerations experienced by a proof mass in a single axis. The most common technologies used to construct an accelerometer are based on piezoelectric materials or microelectromechanical systems (MEMS) devices. In a piezoelectric accelerometer, a piezoelectric material is sandwiched between two proof masses or plates. The forces acting on the masses compress or expand the piezoelectric material generating an electrical potential, which can be amplified and filtered to produce an analog voltage output. In one form of a MEMS accelerometer, such as that shown in Figure 4.10, a thin beam deflects in response to accelerations and thereby changes the capacitance between itself and a fixed beam. A large number of these structures can be machined on an integrated circuit for greater accuracy and to cover multiple axes. A typical MEMS accelerometer has a bias error of 10–100 mg, which varies with temperature.

Angular rates around the vehicle axes can be measured using a rate gyroscope. Compact and inexpensive varieties are based on MEMS technologies. One implementation involves a vibrating piezoelectric structure. Angular rotation induces an out of plane motion through Coriolis forces, as shown in Figure 4.11. Other implementations involve a tuning fork structure where two proof masses oscillate in opposition resonance and out of plane motion is measured, or an oscillating resonant ring, sometimes called a wine glass resonator, in which the nodal positions indicate rotational rates.

Higher-accuracy measurements, at significantly higher cost, can be achieved using ring laser gyroscopes or fiber-optic gyroscopes. In both cases, two light beams travel in opposite directions around a closed path. Angular rotations induce a difference in path length, and the phase shift induced by the difference in path length can be measured by interference fringe patterns as in a Sagnac interferometer, as shown in Figure 4.12. The bias error of a MEMS gyroscope is on the order of 1–10 degrees/second, and for an FOG is generally less than 20 degrees/hour.

Often, the accelerometers and rate gyroscopes are integrated into a single 6 degree of freedom sensor known as an inertial measurement unit (IMU), such as that shown in Figure 4.13. With internal processing, the angular orientations can

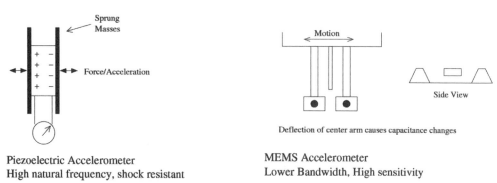

Piezoelectric Accelerometer
High natural frequency, shock resistant

MEMS Accelerometer
Lower Bandwidth, High sensitivity

(a)

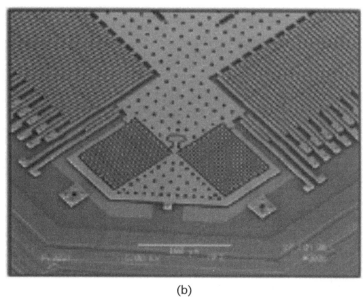

(b)

Figure 4.10   (a) Piezo and MEMS accelerometer and (b) ADXL202 micrograph. (Figure courtesy of Analog Devices.)

be estimated and used to project the accelerations and angular rates into a fixed, ground referenced coordinate system. An example of such a device is described in Table 4.1.

4.2.4   Magnetic Compass (Magnetometer)

One approach to measuring the absolute (yaw) angle of a vehicle is to use an electronic compass that measures the magnetic field of the Earth. Normally this would be accomplished using two or three magnetic sensors arranged along orthogonal axes. The three common sensing technologies in modern use are the fluxgate magnetometer, the Hall effect sensor, and the magnetoresistive magnetometer. In a fluxgate device, two coils surround a core of highly magnetically permeable material, in effect forming a transformer. One coil is fed an oscillating voltage that can saturate the core and this induces a current in the second coil. In the presence of an external magnetic field along the axis of the core, the current in the second coil will be asymmetric with the current in the first coil, and this difference can be measured and

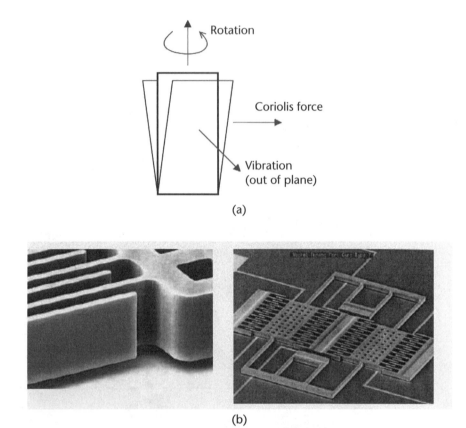

Figure 4.11   (a, b) MEMS gyroscope micrograph. (Figure courtesy of Analog Devices.)

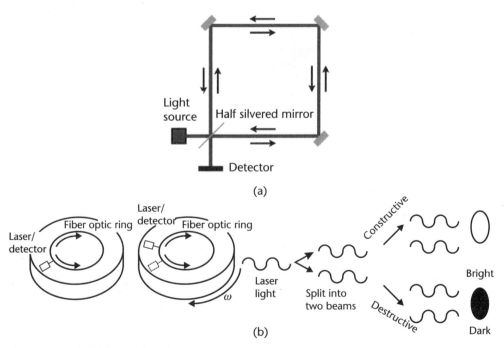

Figure 4.12   (a, b) Sagnac interferometers.

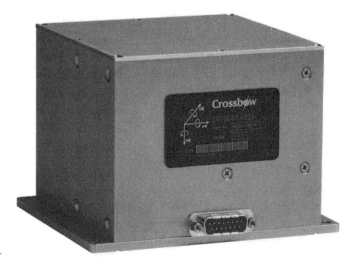

Figure 4.13   Commercial IMU hardware and performance specifications. (Figure courtesy of Memsic.)

Table 4.1   Sample 6 DOF IMU Performance Specifications

Classifications	Items	Performance
	Update rate	> 100 Hz
Attitude	Range: roll, pitch	±180°, ±90°
	Static Accuracy	< ±0.5°
	Resolution	< 0.1°
Angular rate	Range: roll, pitch, yaw	±200°
	Bias: roll, pitch, yaw	< 20°/hr (constant temperature)
	Scale factor accuracy	< 2%
	Nonlinearity	< 1% *FS*
	Resolution	< 0.025°/s
Acceleration	Range: X/Y/Z	±2g
	Bias: X/Y/Z	< 8.5 mg
	Scale factor accuracy	< ±1%
	Nonlinearity	< ±1% *FS*
	Resolution	< ±1 mg

used to estimate the strength of the external magnetic field. The Hall effect sensor is based on the principle that, as charges (electrons or holes) forming a current flow through a conductor, they travel in generally a straight line (excluding collisions), but in the presence of a magnetic field perpendicular to the direction of current flow they travel in a curved path leading to an accumulation of charge on the sides of the conductive material and thus producing an electric potential. Finally, one can use anisotropic magnetoresistive alloys as one leg of a Whetstone bridge. The resistance changes in relation to the applied magnetic field, and this causes a voltage imbalance in the bridge.

Regardless of the technology, interpreting the magnetometer measurements depends on the orientation of the magnetometers relative to the surface of the effort. This can be compensated by utilizing a tilt sensor, either fluidic or accelerometer based, to measure the roll and pitch of the compass. Alternatively, one may use three orthogonal magnetometers, although the tilt sensor approach is generally understood to perform better. Of course, measurements of absolute roll and pitch, even with the low frequency response common in tilt sensors, are also useful in an autonomous vehicle and can be provided as additional outputs.

The magnetic field of the Earth is very weak, so the device must be calibrated to compensate for the presence of external magnetic fields and materials that cause local distortions in the Earth's magnetic field. One class of distortions, usually called a *hard iron effect*, is produced by a nearby magnetized object. In this case, the distortion is fixed regardless of the actual yaw angle of the compass and vehicle, so the effect is to induce a constant offset along the axis of each magnetometer. The other common class of distortion, usually called a *soft iron effect*, is caused by the presence of magnetically permeable materials near the compass. In this case, the distortions in the magnetic field of the Earth are a function of the orientation of the vehicle and generally take the form of different scaling constants along each axis. Assuming the compass is rigidly mounted and the vehicle and its contents do not significantly change, one can perform a calibration procedure, generally involving rotating the compass and vehicle through a set of complete circles, to measure and record the disturbances.

These sensors are also quite sensitive to electromagnetic interference, which can arise from computers, radio transceivers, power electronics, and the internal combustion engine. The author's experience is that the accuracy available in a moving automobile with an internal combustion engine is limited due to the presence of electromagnetic interference and roll and pitch variations at a frequency higher than can be measured and compensated by the compass. Performance is better in an on-road application than an off-road application, and if location and mounting is unconstrained one can arrange to place the sensor in a more suitable area or provide some EMI shielding. Of course, the sensor can be used intelligently in a sensor fusion algorithm that understands the conditions under which the sensor is most reliable (for example, when the vehicle is stationary).

Finally, we note that magnetic north is generally not the same as true north, and the difference, called magnetic declination, varies depending on one's position on the Earth. For example, the current magnetic declination in Columbus, Ohio, is almost 7° west, whereas in Denver, Colorado, it is almost 8° east. It also changes over time, although quite slowly (on the order of 0.05°/year).

4.3 External World Sensing

A number of different sensors have been developed for sensing the external environment of an autonomous vehicle. Many have been developed initially for safety warning or safety augmentation systems that are now being deployed on some high-end vehicles. These include radar sensors, scanning laser range finders, known as light detection and ranging (LIDAR) or sometimes laser detection and ranging (LA-

DAR) in military circumstances, single camera or stereo camera image processing systems, and ultrasonic rangefinders.

To begin, consider Figure 4.14, which illustrates a number of currently available or soon to be available applications for external environment sensing.

4.3.1   Radar

Radar is a popular active sensing technology for road vehicles used in sensing both near and far obstacles. A radar system tends to be designed based on the desired safety or control function. For example, for a fixed, usually regulated output power, there is a general trade-off between field of view and range. For applications like imminent crash detection and mitigation, lane change, and backup safety systems, a shorter range but wider field of view is desired, perhaps on the order of 30 meters of range with a 65–70° wide field of view. For applications like advanced cruise control and crash avoidance, a longer range but narrower field of view is required, perhaps on the order of a 120-meter range and a 10–15° field of view. Two examples of commercial automotive radar systems are shown in Figure 4.15.

Radars are popular choices because they are robust mechanically and operate effectively under a wide range of environmental conditions. They are generally unaffected by ambient lighting or the presence of rain, snow, fog, or dust. They generally provide range and azimuth measurements as well as range rates. They also tend to be available at a lower cost relative to other active sensors such as LIDAR, although their measurements, in particular azimuth angle, are less precise. Their price is decreasing steadily as they are being produced in larger quantities.

Radar sensors have been built to operate on a number of frequencies, but vehicular applications appear to be standardizing on 24 GHz and 77 GHz, with a few very short-range sensors operating at 5.8 GHz or 10.5 GHz. They may be based on a pulsed Doppler or one of many continuous wave modulations. Early radar systems, for example the Delphi ACC radar system, employed a mechanically rotated antenna to generate azimuth angle measurements, but this is difficult to manufacture robustly and inexpensively. Most modern radars are based on a multielement

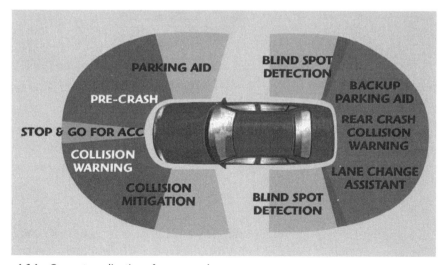

Figure 4.14   Current applications for sensor data.

(a)

(b)

Figure 4.15 (a) MaCOM SRS 24-GHz UWB radar. (b) Tyco long-range 77-GHz radar. (Figures courtesy of Cobham Sensors.)

patch antenna and use DSP-based antenna pattern beam-forming techniques to measure azimuth angle.

4.3.2 LIDAR

A scanning laser range finder system, or LIDAR, is a popular system for obstacle detection. A pulsed beam of light, usually from an infrared laser diode, is reflected from a rotating mirror. Any nonabsorbing object or surface will reflect part of that light back to the LIDAR, which can then measure the time of flight to produce range distance measurements at multiple azimuth angles. The basic idea is shown in Figure 4.16.

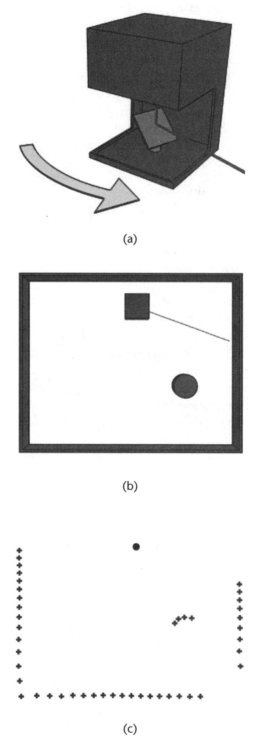

(a)

(b)

(c)

Figure 4.16   (a–c) LIDAR scans.

LIDAR sensors can produce very precise measurements. Higher-end sensors can produce multiple distance measurements per laser pulse, which can be helpful when trying to see through dust, rain, mostly transparent surface (such as glass windows), and porous objects such as wire fences. Some recent sensors can scan in multiple planes in order to provide a semi-three-dimensional view of the external world. This can be quite useful since, with a single plane, it is difficult to distinguish between ground strikes and objects of interest as the vehicle pitches and rolls. Most also provide some measure of the intensity of the reflected pulse. Three representative commercially available LIDAR sensors are shown in Figure 4.17, and their functional specifications are compared in Table 4.2. LIDAR scanners are still quite expensive, and there are some questions about the long-term reliability of the mechanical scanning mechanisms. While they tend to be heavily used in research applications, they have not seen wide use in automotive OEM safety applications.

4.3.3 Image Processing Sensors

Vision and image processing sensors have been heavily used in automated vehicle research, and have been deployed in some safety applications. In general terms, single camera systems are often used for lane marker or lane edge detection [3, 4] and for basic, low-accuracy object detection and localization, and are currently applied in lane departure warning systems and a few forward collision warning systems. Researchers have also developed road sign reading applications. Multiple camera systems, for example, stereo vision systems, can provide a depth map for objects in the world and can be used for obstacle detection. Although, to date, they have been applied mostly in research applications, a few OEM automotive products are beginning to appear.

Their primary advantage is their use of low-cost, off-the-shelf components and that their implementation is almost entirely in software, which makes them quite amenable to research development and testing. Their primary disadvantage is that they are almost always implemented as passive sensors, and thus must cope with the full range of ambient and uncontrolled conditions, including lighting, shadowing and reflection, and atmospheric weather, dust, and smoke.

Table 4.2   Specifications of Several LIDAR Sensors

	SICK LMS291	Ibeo LUX	Velodyne HD-32E
Number of layers	1	2 (110°), 4 (85°)	32
Horizontal field of view	180°	110°	360°
Vertical field of view	N/A	3.2°	−10° to +30°
Update rate (Hz)	Up to 75 Hz	12.5–50 Hz	5–20 Hz
Range (meters)	80	200m	100m
Range (10% reflectivity)	30m	50m	50m
Range accuracy	4.5 cm	10 cm	2 cm
Horizontal accuracy		Up to 0.125°	
Vertical accuracy	N/A	0.8°	0.08°
Horizontal resolution	0.25°, 0.5°, 1.0°	4 cm	0.08°
Vertical resolution	N/A	Varies	1.25°
Interface	RS232/RS422	Ethernet	Ethernet

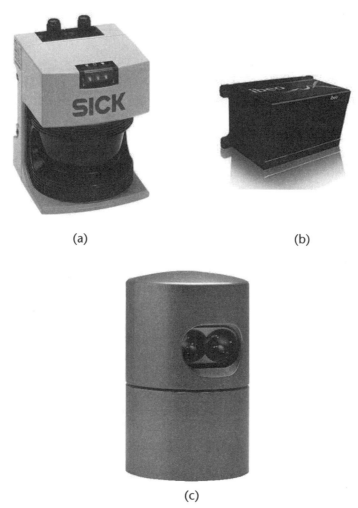

(a)   (b)

(c)

Figure 4.17   (a–c) LIDAR sensors (SICK LMS291, Ibeo Lux, and Velodyne HD-32E). [(a) Image courtesy of SICK AG, (b) image courtesy of Ibeo Automotive Systems GmbH, and (c) image courtesy of
Velodyne Lidar Inc.]

4.3.3.1   Single Camera Lane Marker Sensor

The extraction of lane markers from video images is a common function that can
be applied in an autonomous vehicle, since once the lane markers are located the
lateral position of the vehicle can be determined and this information can be used
in a steering control algorithm for lane keeping. This approach allows autonomous
operation without relying on the availability of precise vehicle position or accurate
and up-to-date road and lane maps.

Many different approaches have been applied to this problem. They include
various region-based approaches, which involve either color or contrast/brightness
segmentation of the image into regions of similar characteristics, voting procedures
such as edge detection followed by a Hough transform, and feature-based extraction systems that utilize voting or RANSAC techniques. Many algorithms begin
with an inverse perspective transform, which converts the camera plane image to a

rectified, bird's-eye view. A summary of the lane detection problem and other image processing problems and solution techniques relevant to on-road vehicles can be found in [5].

As a representative example of a sensor suitable for lane marker location, we present a system for extracting lane marker information from image data in a form suitable for use in the automated steering of moving vehicles developed at The Ohio State University. The algorithm [3] was designed with speed and simplicity in mind. It assumes a flat, dark roadway with light-colored lane markers, either solid or dashed, painted on it. The system was initially developed in 1995, and has been tested in various autonomous vehicle control research projects, including steering control at highway speeds at the 1997 AHS Technical Feasibility Demonstration in San Diego.

The basic algorithm is as follows. First, we implement an adaptive adjustment of the black and white brightness levels in the frame grabber in order to maximize the dynamic range (contrast) in the region of interest under varying lighting conditions. Then, using the history of located lane markers in previous image frames, information about lane markers that have already been located in this frame, and geometric properties of the ground to image plane projection of lane markers (for example the convergence of parallel lines at infinity and the known or estimated width of the lane), we identify regions of interest at a number of look ahead distances.

For each region of interest we apply a matched filter, tuned to the average width of a lane marker at the given distance ahead, to the pixel brightness values across the image and store this information in a corresponding vector. We extract from each vector those candidate lane marker points (bright spots) that pass a number of statistical hypothesis tests based on the standard deviation and absolute magnitude of the candidate points and the minimum overall standard deviation of the nearby area of the image.

Assuming that the lane markers on the road are parallel, we can fit a low-order polynomial to the computed location of the middle of the lane at all look ahead distances for which lane markers were identified. This curve allows us to estimate a lateral offset distance between the center of the lane and longitudinal axis of the vehicle at any look ahead distance. We can also estimate the curvature of the lane.

By using curve fitting and looking well ahead, this algorithm can handle broken or dashed lines. Also, by adaptively estimating the width of the lanes or by entering the lane width directly into the algorithm, the software can estimate the position of the vehicle in the lane from only one lane marker.

It was originally implemented on a TI TMS320C30 DSP system using a low-cost monochrome CCD camera. The camera was mounted at the lateral center of the vehicle as high as possible in order to obtain the best viewing angle. An example of its operation is shown in Figure 4.18. We estimate that it logged over 1,500 miles both on I-15 and on a demonstration site with tight (150-foot radius) curves and, with the appropriate tuning of parameters and filter coefficients, was tested and found to work well on a number of different road surfaces (new asphalt, worn asphalt, and concrete) and under different lighting conditions.

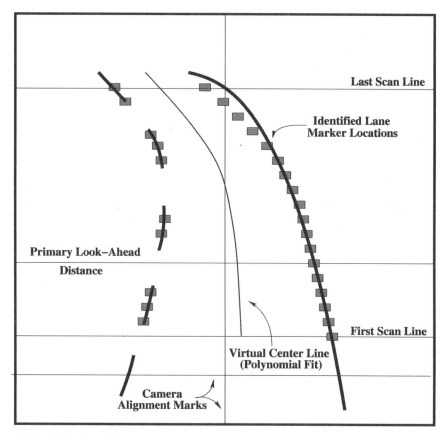

Figure 4.18 Vision-based lane marker extraction.

4.3.3.2 Stereo Vision Obstacle Sensor

Another application of image processing is the detection of obstacles. This can be accomplished using a stereo vision system, or in some cases using an optical flow approach.

It is possible to extract a depth or height map of objects in the world using two (or more) cameras located such that they view the world from two slightly different perspectives. Usually they are mounted horizontally with some separation that is related to the desired depth of field and resolution. The principle of stereo image processing is as follows. Consider the idealized situation shown in Figure 4.19, in which it is shown that all the points in the real three-dimensional world that lie along a single line passing through the optical center of a camera, for example, P_1 and P_2 as shown, project onto the same point P in the image plane. As expected, the depth or distance information about P_1 and P_2 are lost.

Using two images taken from different perspectives, it is possible to recover the depth information. Consider the simplified example shown in Figure 4.20. Two cameras are mounted such that their image planes are coplanar as shown. A point P in the real world is projected onto the left and right camera image planes as P_l and P_r. X_l and X_r are the distances from the image plane center to the projected points. O_l and O_r are the optical centers of the left and right cameras. B is the

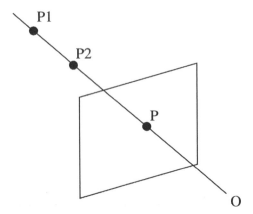

Figure 4.19 Projections of points onto the image plane.

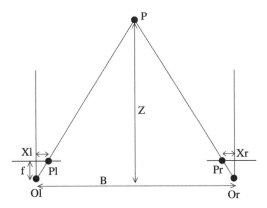

Figure 4.20 Planar example of the recovery of depth using stereo images.

baseline length, and f is the focal length of the idealized (pinhole) camera. Then the disparity is given as

$$d = x_r - x_l$$

and the depth, derived by simple triangulation, is given by

$$Z = f\frac{B}{d}$$

Even when expanded to a two-dimensional image plane the computation is not difficult. What is difficult, however, is identifying the points P_l and P_r in the images of the two cameras that correspond to a single point P in the real world using only the two images and no knowledge of the real world. Many techniques have been suggested. Most are based on identifying some recognizable features in the image, for example corners, and finding the correct matches on each image.

We note that the density of the resulting disparity or depth map is a function of the number of correspondence points that can be identified and the distribution of those points in the real world. We also note that depth Z is inversely proportional to disparity, so that the resolution of the depth map decreases for objects further from the camera due to the projection of a larger area onto a single pixel as distance increases.

The following four steps comprise the general stereo vision algorithm:

1. Correction of camera and lens-based image distortions;
2. Rectification of both images to a common reference plane;
3. Identification of features (points, corners, patterns, and so forth) visible in both images and measuring the displacement in order to compute a disparity map;
4. Using camera and mounting parameters to convert the disparity map to a height or depth map.

4.3.4   Cooperative Infrastructure Technologies

Over the last 20 years a number of technologies that involve adding components or electronic systems to the roadway to enhance on-vehicle sensing systems have been developed to assist an autonomous vehicle in following a roadway. Most prominent are:

- Magnets placed in the middle of the lane on a roadway;
- Radar reflective stripe placed on the roadway;
- RF tags placed on the side of a road at regular intervals.

We briefly discuss the radar reflective stripe technology developed by The Ohio State University [6]. The radar sensor measures lateral position by sensing backscattered energy from a frequency selective surface constructed as lane striping and mounted in the center of the lane. The conceptual basis for the radar lane tracking system is shown in Figure 4.21. The radar reflective surface is designed such that radar energy at a particular frequency is reflected back toward the transmitting antenna at a specific elevation angle. Thus, by varying the frequency of the radar signal we can vary the look ahead distance of the sensor.

The radar chirps between 10 and 11 GHz over a 5-millisecond period, transmitting the radar signal from a centrally located antenna cone. Two antenna cones, separated by approximately 14 inches, receive the reflected radar energy. The received signal is downconverted into the audio range by mixing with the transmit signal. The lateral offset of the vehicle is found as a function of the amplitude of the downconverted left and right channel returns at a radar frequency corresponding to a particular look ahead distance.

In addition, the peak energy in the downconverted signal appears at a frequency that is a function of the distance from the vehicle to an object ahead. It is thus possible to extract the distance to an object ahead of the automated vehicle using the radar hardware already in place for lateral sensing.

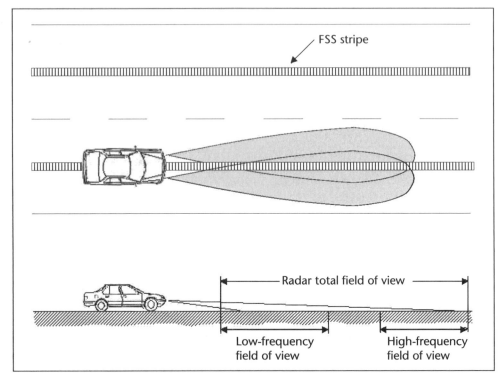

Figure 4.21   The OSU radar reflective stripe sensor.

An operational example of this technology is shown in Figure 4.22, in which the radar system for lateral sensing and control is mounted behind the front bumper shroud of each vehicle and the radar reflective stripe and can be seen in the center of each lane.

Other cooperative infrastructure technologies include wireless infrastructure-to-vehicle communication from traffic signals, road maintenance equipment, and

Figure 4.22   One autonomous car passing another at TRC proving grounds. The radar reflective stripe can be seen on both lanes.

possibly road or traffic information systems, active pedestrian crosswalk or intersection boundary markers, and GPS enhancement systems.

4.4 Estimation

Even under the best of circumstances sensors provide noisy output measurements.
In autonomous vehicles, measurements from sensors of varying reliability are used
to ascertain the location and velocity of the car both with respect to the roadway
or locations on a map and with respect to obstacles or other vehicles. In order to
use sensor measurements for control purposes the noise components may need to
be eliminated, measurements may need to be transformed to match variables, and
measurements from multiple sensors may need to be fused. Fusion of data may
be required because a single sensor or sensor reading is insufficient to provide the
needed information, for example the shape of an obstacle or open pathway, or it
may be needed to provide redundancy and reduce uncertainty.

4.4.1 An Introduction to the Kalman Filter

The Kalman filter is used to generate an estimate of the state variables of a system
from noisy measurements given a known model of the system. It is an optimal linear
filter for systems with Gaussian noise, and though that condition may not be true in
practice the Kalman filter is still often applied with good results.

We consider the linear, discrete time, time-varying system modeled as

$$x_k = F_k x_{k-1} + B_k u_k + w_k$$

where

F_k is the state transition matrix;

B_k is the control-input matrix;

w_k is the process noise, which is assumed to be a zero mean normal distribution
with covariance matrix Q_k;

and

$$w_k \sim N(0, Q_k)$$

The output of the process, that is, the measured variables, is

$$z_k = H_k x_k + v_k$$

where

H_k is the output matrix modeling the linear combination of states that are
measured;

v_k is the observation noise, also assumed to be zero mean Gaussian, with covariance R_k;

and

$$v_k \sim N(0, R_k)$$

The initial state and the noise vectors at each time instant are mutually independent.

The Kalman filter estimates the state of a system based only on the last estimate and the most recent set of measurements available, and therefore is a recursive filter.

The state of the filter is represented by two variables:

$\hat{x}_{k|k}$, the estimate of the state at time k;

$P_{k|k}$, the error covariance matrix (a measure of the estimated accuracy of the state estimate).

The Kalman filter has two distinct phases. The *prediction phase* is where the state is estimated for the next time instant based on the previous estimate, but with no new measurements.

$$\hat{x}_{k|k-1} = F_k \hat{x}_{k-1|k-1} + B_k u_k$$

Note that this is just using the known state equations with no concept of noise.

The *update phase* is where the predicted state is corrected when a new set of measurements is available:

$$\hat{x}_{k|k} = \hat{x}_{k|k-1} + K_k \tilde{y}_k$$

where

$$\tilde{y}_k = z_k - H_k \hat{x}_{k|k-1}$$

shows the error we have made in estimating the output/measurement. It is sometimes called the *innovation* or *residual*. The matrix **K** is called the *Kalman gain*.

The Kalman gain is calculated through a series of equations involving the predicted state estimate covariance

$$P_{k|k-1} = F_k P_{k-1|k-1} F_k^T + Q_k$$

the covariance of the residual

$$S_k = H_k P_{k|k-1} H_k^T + R_k$$

the calculation of the optimal Kalman gain

$$K_k = P_{k|k-1}H_k^T S_k^{-1}$$

and the update of the state estimate covariance

$$P_{k|k} = (I - K_k H_k)P_{k|k-1}$$

Usually one assumes that the estimated distributions for $\hat{x}_{0|0}$ and $P_{0|0}$ would initially be known. Sometimes the values of $\hat{x}_{0|0}$ is known exactly and then $P_{0|0} = 0$.

Although one generally conceives of the Kalman filter operating in a cycle of predict, update, and repeat, it is possible to skip the update step if no new measurements are available or to execute multiple update steps if multiple measurements are available or if multiple **H** matrices are used because of the presence of different sensors possibly updating at different rates.

We have only presented a brief summary of the Kalman filter in the linear systems case. There are a number of the variants or expansions of the Kalman filter, the best known being the extended Kalman filter, in which the state transition and the output or measurement functions are nonlinear and the covariance matrices and Kalman gains are computed using a linearization of the system, known as the Jacobian, around the current state. The reader is referred to [7, 8] for a more thorough treatment of the Kalman filter and other estimation techniques.

4.4.2   Example

As a very simple example, consider that one might want to estimate the speed of a target vehicle detected by some sensor that measures the position of the vehicle over time. This could be a LIDAR sensor or an image processing sensor. For simplicity, we assume the sensor is stationary, for example it could be mounted on the side of the road or above the roadway. We also assume that the vehicles passing the sensor are modeled with point mass dynamics, without friction, and that they are experiencing an unknown random acceleration (or deceleration). In this case, the **F, H, R,** and **Q** matrices are constant and so we drop their time indices.

The model of the vehicle consists of two states, the position and velocity of the vehicle at a given time

$$x_k = \begin{bmatrix} x \\ \dot{x} \end{bmatrix}$$

We assume that between each sampling timestep, for example, the $(k-1)_{th}$ and k_{th} time step, each vehicle has a random change in its acceleration a_k that is normally distributed with mean 0 and standard deviation σ_a which is treated as a process noise input. The point mass dynamics with no friction is

$$x_k = Fx_{k-1} + Ga_k$$

where

$$F = \begin{bmatrix} 1 & \Delta t \\ 0 & 0 \end{bmatrix}$$

and

$$G = \begin{bmatrix} \dfrac{\Delta t^2}{2} \\ \Delta t \end{bmatrix}$$

We find that, since σ_a is a scalar variable,

$$Q = \mathrm{cov}(Ga) = E\left[(Ga)(Ga)^T\right] = GE\left[a^2\right]G^T = G\left[\sigma_a^2\right]G^T = \sigma_a^2 GG^T$$

The position measurement is

$$z_k = Hx_k + v_k$$

where

$$H = \begin{bmatrix} 1 & 0 \end{bmatrix}$$

and the measurement or sensor noise is also normally distributed, with mean 0 and standard deviation σ_z, that is:

$$R = E\left[v_k v_k^T\right] = \left[\sigma_z^2\right]$$

If the system is triggered by a vehicle passing by it, we may choose to assume that the initial state of the vehicle is known perfectly and given by

$$\hat{x}_{0|0} = \begin{bmatrix} 0 \\ 0 \end{bmatrix}$$

with a zero covariance matrix:

$$P_{0|0} = \begin{bmatrix} 0 & 0 \\ 0 & 0 \end{bmatrix}$$

Having defined the model and the various parameters for the Kalman filter, we can now select an appropriate time step, possible equal to the update rate of the sensor, and implement the Kalman filter computations to estimate the vehicle state, including the unknown speed.

4.4.3   Another Example of Kalman Filters: Vehicle Tracking for Crash Avoidance

In this example we consider using LIDAR data to track a vehicle approaching an intersection, as shown in Figure 4.23. In this scenario, a host vehicle is equipped with LIDAR sensor(s) and desires to estimate an approaching vehicle's x and y position, velocity, and acceleration. The LIDAR sensor provides distance measurements r to the closest object at varying azimuth angles.

To ensure filter convergence it is necessary to receive measurements from a stable point on the other vehicle. Thus, a corner finding algorithm is used to estimate the other vehicle's corner location based on the LIDAR returns. Unfortunately, LIDAR returns from the other vehicle may not always fit a corner model. However, a pseudo-corner can be estimated if the returns fit a linear model. Figure 4.24 depicts the LIDAR returns (marked as Xs) from the other vehicle and the estimated corner location for one time step.

It is possible to track the other vehicle in polar coordinates using a nonlinear filter such as the extended or unscented Kalman filter. However, doing so increases the complexity of the filter equations and requires proper handling of the angle periodicity. Instead, the polar measurements can be transformed to Cartesian measurements and a linear filter can be used to track the other vehicle.

The host vehicle tracks the other vehicle via a six-dimensional state vector containing the x and y position, velocity, and acceleration. Moreover, a constant acceleration state transition matrix is used to update the state vector with the assumption that a random perturbation acts on the velocity of the other vehicle.

If T is the sampling period, the Kalman filter equations become:

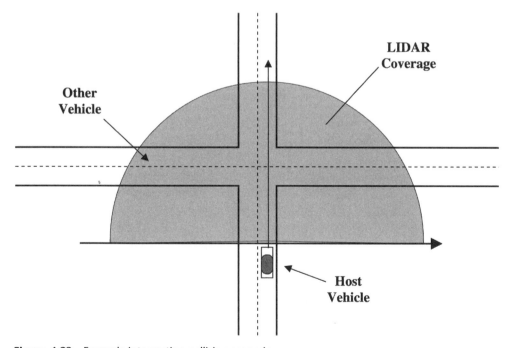

Figure 4.23   Example intersection collision scenario.

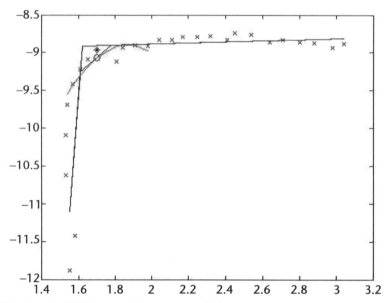

Figure 4.24   Sample LIDAR returns from target vehicle.

$$x_{k+1} = Fx_k + Dw_k$$
$$z_{r_k} = r_k + v_{r_k}$$
$$z_{\theta_k} = \theta_k + v_{\theta_k}$$

where

$$w_k \sim N(0, Q_k), \quad v_k \sim N(0, R_k)$$

$$x_k = \begin{bmatrix} x_{pos} \\ x_{vel} \\ x_{acc} \\ y_{pos} \\ y_{vel} \\ y_{acc} \end{bmatrix}, \quad D = \begin{bmatrix} 0 & 0 \\ 1 & 0 \\ 0 & 0 \\ 0 & 0 \\ 0 & 1 \\ 0 & 0 \end{bmatrix}, \quad F = \begin{bmatrix} 1 & T & 0.5T^2 & 0 & 0 & 0 \\ 0 & 1 & T & 0 & 0 & 0 \\ 0 & 0 & 1 & 0 & 0 & 0 \\ 0 & 0 & 0 & 1 & T & 0.5T^2 \\ 0 & 0 & 0 & 0 & 1 & T \\ 0 & 0 & 0 & 0 & 0 & 1 \end{bmatrix}, \quad C = \begin{bmatrix} 1 & 0 & 0 & 0 & 0 & 0 \\ 0 & 0 & 0 & 1 & 0 & 0 \end{bmatrix}$$

The prediction step consists of:

State: $\overline{x_k} = Fx_k$
State covariance: $\overline{P_k} = FP_kF^T + DQ_kD^T$

The update step consists of:

Measurement covariant: $P_{zz} = C\overline{P_k}C^T + R_k$
State/measurement covariance: $P_{xz} = \overline{P_k}C$
Kalman gain: $K_k = P_{xz}\left(P_{zz}\right)^{-1}$

$$\text{Measurement residual: } y_k = \left(\begin{bmatrix} r_k \cos(\theta_k) \\ r_k \sin(\theta_k) \end{bmatrix} - C\overline{x_k} \right)$$

$$\text{State: } x_{k+1} = \overline{x_k} + K_k y_k$$

$$\text{State covariance: } P_{k+1} = (I_6 - K_k C)\overline{P_k}$$

Figure 4.25 shows results from an experimental test where the host vehicle travels along the line $y = x$ with a final velocity of 8.0619 m/s and tracks the other vehicle that is traveling along the line $y = -x$ with a final velocity of 8.9762 m/s.

4.5   Sensor Fusion

In the previous sections we have looked at the various types of sensors that might be used in an autonomous vehicle. However, as we shall see, usually a vehicle is fitted with multiple sensors and technologies, and the outputs of these individual sensors must be combined to produce a final overall view of the world. The primary justifications for this approach are:

- Different sensors and sensor technologies have different perceptive abilities for the environment.
- Different or multiple sensors have different fields of view.
- A single sensor does not cover the application's required field of view.

However, when multiple sensors or sensor technologies are deployed, there is always the chance of conflict between the sensor measurements. For example, the same object in the environment may be reported at two different positions (i.e., 30 meters versus 28 meters), or one sensor may detect an object and another may not. These conflicts may arise because of:

- Different perceptive characteristics of each sensor: for example, range, accuracy, field of view, sensitivity, or response to environmental factors;
- Actual changes over time in the environment;
- Faulty sensor element, electronics, noise, or false data input;
- Thresholds or other differences between processing algorithms.

There are various techniques to deal with these conflicts, including tracking and filtering (i.e., extended Kalman filter), confidence and hypothesis testing approaches, voting schemes, and evidence-based decision theory.

4.5.1   Vehicle Localization (Position and Orientation)

A key element of autonomous vehicle technology is vehicle localization. All aspects of the system, from sensor processing and fusion to navigation and behavioral decision making to low-level lateral and longitudinal control, require accurate vehicle position, velocity, and vehicle heading, pitch, and roll information at a fairly high update rate. Providing this information requires the use of multiple sensors, possibly

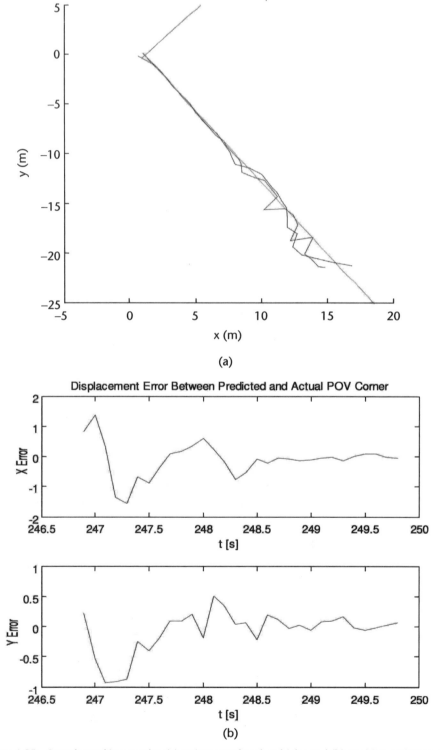

Figure 4.25   Sample tracking results: (a) trajectory of each vehicle; and (b) position estimate errors.

including one or more Global Positioning System (GPS) receivers augmented with some correction service, for example, Omnistar HP wide-area differential corrections, inertial measurement units (IMU), and dead reckoning sensors (wheel speeds, transmission gear and speeds, throttle, brake, and steering wheel position) provided on the vehicle, and a validation system to eliminate sensor errors, especially GPS-related step-change events caused by changes in differential correction status or the visible satellite constellation. To account for sensor errors, noise, and the different update rates of each sensor, an extended Kalman filter is applied to generate the required state measurements [9, 10].

The reasons for fusing these sensors are:

- *Accuracy*: IMU integration can lead to the unbounded growth of position error, even with the smallest amount of error or bias in its measurements. This gives rise to the need for an augmentation of the measurements by external sources to periodically correct the errors. GPS can provide this, since it provides a bounded measurement error with accuracy estimates.

- *Data availability*: GPS is a line-of-sight radio navigation system, and therefore GPS measurements are subject to signal outages, interference, and jamming, whereas an IMU is a self-contained, nonjammable system that is completely independent of the surrounding environment, and hence virtually immune to external disturbances. Therefore, an IMU can continuously provide navigation information when GPS experiences short-term loss of its signals. Similarly, dead reckoning sensors are internal to the vehicle.

Figure 4.26 shows one configuration of a vehicle localization sensor fusion system. This example is often called a loosely coupled system because the final outputs of the individual sensors are fused.

Several commercial manufacturers also provide tightly coupled systems, in which low-level raw data, for example GPS pseudorange and Doppler measurements, are directly fused with IMU and dead reckoning data.

4.5.2   External Environment Sensing

There are a number of issues when designing the external environment sensing suite for an autonomous vehicle. The system objectives and requirements are obviously a direct influence in the design process. For example, highway driving is a much more structured environment than off-road driving, and that structure can be exploited in sensor and control system design. On the other hand, the urban environment, which may consist of irregular and changing road networks and vehicles and pedestrians behaving unpredictably, is much less structured than highway driving and may require significantly more sensing capability.

Robustness and safe performance of hardware and software are obviously required for both off-road driving and on-road production automobiles. Certain problems are mitigated when a system is designed for a passenger vehicle environment. Nevertheless, business concerns including marketing and liability protection demand a high level of robustness. Redundancy in sensing modalities is also a highly desired feature, especially in a less structured, more uncertain environment,

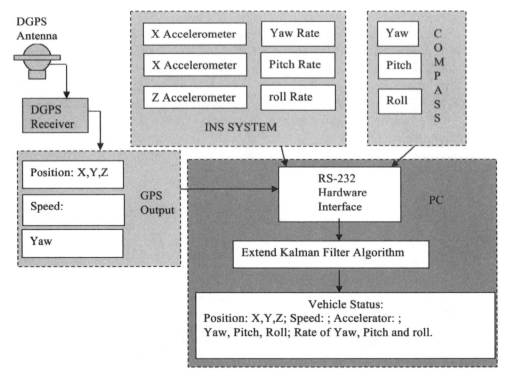

Figure 4.26   Vehicle localization fusion system.

but the cost of many of the sensors used on research vehicles would be prohibitive for a commercial passenger vehicle application.

The level of autonomy with respect to the human driver is also a significant design issue in a passenger vehicle system. The degree to which the driver is to be part of the sensing and control loop is a design decision driven both by technical and nontechnical (i.e., marketing and legal) considerations. Driver attention and situation awareness, human-machine interface and driver workload considerations, and driver state considerations must be considered.

The availability of a priori data, whether from the various terrain and digital elevation map and satellite imagery datasets that might be considered useful for fully autonomous off-road route planning and navigation, to the road map datasets, traffic condition reports, and road maintenance activity schedules that would be useful for passenger vehicle automation or driving enhancements is also a significant issue. Error correction and real-time updates of a priori data sets are obviously useful and necessary for future vehicle systems.

Finally, vehicle to vehicle and vehicle to infrastructure communication capabilities will almost certainly be involved in future vehicle systems, opening the potential of traffic cooperation and facilitating everything from navigation and routing systems to traffic control systems to driver warning and collision avoidance systems.

A general architecture for an autonomous vehicle sensor system is shown in Figure 4.27. There are some distinctions in considering sensing requirements for urban versus off-road applications. We list some noteworthy items here:

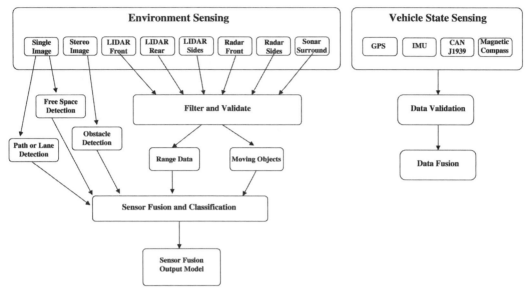

Figure 4.27 Representative vehicle sensing system.

- For pure highway applications, a very restricted approach may be usable. Since there are a number of distinct issues of concern, for example lane edge location, and static or moving obstacles in relevant locations, one can simply fuse data related to specific tasks and not necessarily provide a complete and integrated representation of the world.

- For off-road applications, compensation for vibration and other vertical and rolling motions needs to be done in software or hardware, for example using the IMU and sensor data to specifically generate a ground plane that can be referenced while performing sensor validation and fusion. Sensor adjustments are also required to deal with dust, rain, and changing lighting conditions.

- For domains where there are many moving obstacles (i.e., urban applications), one may need to track individual objects at all times.

- Specific operations (parking, dealing with intersections, entering/exiting highways, and so forth) may use totally separate sensing and sensor architectures tailored to the task.

As will be discussed more fully in the following sections, the sensor coverage must be tailored to the application. The primary consideration for off-road driving is obstacles in front of and immediately beside the vehicle. Inexpensive short range ultrasonic sensors may be sufficient to allow for a small, slow backup maneuver and to provide side sensing in very tight corridors. In an urban scenario, where sensing is required to support sideways maneuvers into alternate lanes at high speeds, u-turns, intersection negotiation, and merging into oncoming traffic, sensing for a significant distance in all directions around the vehicle is required.

In general, we identify two approaches to sensor fusion and the representation of a world model: a grid (cell) or occupancy map approach [11] and a cluster identification and tracking approach [12, 13]. For a grid map approach the sensing architecture for sensor fusion is established by developing a cell-based grid map of the vehicle's surroundings. All external sensors feed into this map with obstacles sensed and related confidence levels, and the grid map provides both the framework for sensor fusion and the representation of the world model. This tends to assume that most features in the environment are stationary and that the environment is fairly unstructured, and applies most obviously to an off-road driving application. For the cluster identification and tracking approach, the sensor fusion algorithm is responsible for fusing the data from multiple sensors into geometric clusters and tracking those clusters, usually by estimated their velocities, accelerations, and possibly trajectories over time and extrapolating estimated motion into the future. In this case the world model may be expressed as a discrete data structure, for example a tree or a linked list representing a graph. This approach is more suited to traffic situations where the environment is highly dynamic.

One can render an occupancy map from the track list fairly easily should this representation be required. Creating a track list from an occupancy map requires carrying out the clustering and segmentation step and thus is more complicated. This is to be expected, as the track list is a higher-level, more idealized representation of the world than an occupancy map.

4.5.3 Occupancy Maps and an Off-Road Vehicle

For an off-road vehicle or robot, there are a great variety of target types that need to be detected and registered, as well as a lack of high-quality a priori information about the environment. The vehicle and its sensors will also be exposed to significant disturbances such as vehicle motions, dust, and debris. The ideal approach is to provide, within the limits of cost, available physical space, electrical power, and computational power, a diversified sensor system with as many different sensor modalities as possible [14].

4.5.3.1 Sensor Suite for an Example Off-Road Vehicle: The OSU ION

The Ohio State University participated in the 2005 DARPA Grand Challenge, an off-road autonomous automated vehicle race held in the desert southwest (Nevada and California) of the United States. The goal of the sensor suite and sensor fusion module was to provide 360° sensor coverage around the vehicle while operating in an entirely unknown environment with unreliable sensors attached to a moving vehicle platform [15]. Budget constraints required that this be accomplished without significantly expanding the existing sensing hardware available to the team. The chosen sensor suite is shown in Figure 4.28. The effective range of each sensor is also indicated.

Three SICK LMS 221-30206 180° scanning laser rangefinders (LIDARs) were mounted at three different heights: the first at 60 cm above the ground and scanning parallel to the vehicle body, the second at 1.1 meters above the ground and scanning in a plane intersecting the ground approximately 30 meters ahead of the vehicle, and the third at 1.68 meters above the ground with the scanning plane intersecting the ground approximately 50 meters ahead of the vehicle. The use of three LIDARs allowed a rough estimate of object height to be computed as the

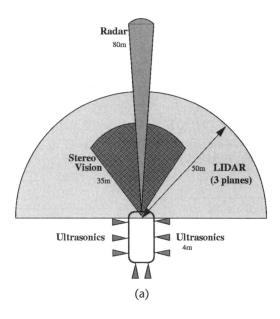

(a)

(b)

Figure 4.28   (a, b) The OSU ION off-road vehicle sensing systems.

vehicle approached an obstacle. A fourth LIDAR, not shown in Figure 4.28(a), was mounted at 1.68 meters above the ground and scanning in a vertical plane. This LIDAR provided an estimate of the ground profile directly ahead of the vehicle, which is crucial for eliminating ground clutter and the effects of vehicle pitching and bouncing motions. An Eaton-Vorad EV300 automotive radar with a 12° scanning azimuth and an 80–100-meter range was also mounted parallel to the vehicle body alongside the lower LIDAR.

A stereo pair of monochrome Firewire cameras and an image processing system, described below, was also installed on the vehicle. It was rigidly mounted in solid housings and included devices for system ventilation and windscreen cleaning. The algorithmic structure of the video sensor platform for ION is depicted in Figure 4.29 [16].

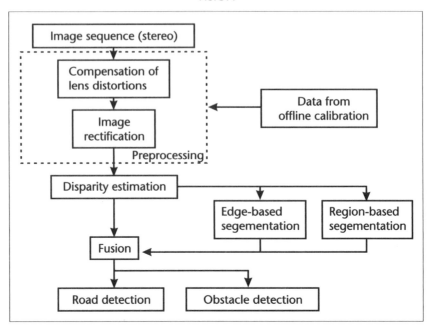

Figure 4.29 Image processing system.

In a preprocessing step, the images acquired from the stereoscopic camera system are warped to compensate for lens distortion and then rectified to compensate for imperfect alignment and coplanarity of the two cameras. In order to achieve a high degree of robustness against variable environmental conditions, a diversity of features is exploited by the subsequent processing step. Disparity, color homogeneity, and orientation are the three primary features computed. The disparity feature allows a fast and robust computation of the ground plane parameters. Similar to estimation in Hough space, the v-disparity technique searches for a linear decreasing disparity along the columns [17]. Disparity is a reliable clue for depth in well-textured regions near the stereo camera. In contrast it is highly unreliable in regions of homogeneous color. Humans possess the ability to interpolate over such regions. We aim to mimic this capability by segmenting a monoscopic image into nonoverlapping regions that include homogeneous colors. Hence, large deviations in color may only occur across region boundaries.

Finally, eight Massa M-5000/95 ultrasonic rangefinders were mounted around the vehicle to provide side sensing for narrow passages (including tunnels and bridges) and rear sensing for the vehicle while driving in reverse. Two additional ultrasonic rangefinders were mounted high on the vehicle and angled downward at approximately 45° to detect drop-offs and cliff faces near the left and right sides of the vehicle.

In addition to the sensors that monitor the external environment, localization and orientation sensors, including a Novatel ProPak-LB-L1/L2 GPS using the Omnistar HP wide-area differential correction service, a Crossbow VG700A fiber-optic based vertical gyroscope, a Honeywell digital compass, and wheel speed sensors on the front and back wheels were installed, validated in real-time, and fused using an extended Kalman filter to provide position, angular orientation, and speed information to both the sensor fusion and vehicle control modules [9, 10].

4.5.3.2 Ground Profile Estimation

In an ideal, static situation, the various sensors provide an accurate map of the objects around the vehicle. However, in many situations it is difficult to decide whether the sensors are returning measurements from the ground or measurements representing an obstacle.

An off-road vehicle rapidly changes orientation as it traverses rough terrain. Bumps and small hills may cause the vehicle to pitch violently, while ruts on either side of the road, as well as steering motions of the vehicle, may induce roll. It is assumed that the sensors are fixed with respect to the vehicle, although strictly speaking the sensor mounts may flex under high dynamic loads. These effects could be mitigated, although not entirely eliminated, using an actuated stabilized sensor mount if available. In any case, estimation of the ground profile is required to eliminate unwanted terrain returns from the raw sensor data.

One approach to this problem would be to use horizontally mounted scanning LIDARs to both detect objects and also identify the ground terrain in front of the vehicle. This technique has been successfully used in on-road vehicles. If the scan of the horizontal LIDARs intersects flat ground in an obstacle-free environment, the measurements should form a line, which reveals the pitch and roll of the vehicle. If this ground line could be detected in a realistic, noisy environment, the vehicle's orientation with respect to the ground could be determined. As a simple example, a candidate linear feature $(\bar{x}, \bar{y}, v)$ can be obtained from the raw data using a least squares linear regression. Sensor returns that do not lie within some threshold distance of the line passing through $(\bar{x}, \bar{y})$ in the direction of the unit vector v are iteratively removed from the matrix such that the accuracy of the discovered linear feature is refined. However, in actual tests such a line fine algorithm failed to identify the ground line in an unstructured, cluttered environment or when the ground ahead was not sufficiently planar.

Another potential approach involves using the pitch and roll measurements from a vertical gyroscope to estimate the height of objects seen by the LIDARs. If height estimates are sufficiently accurate, LIDAR measurements can be filtered to remove ground returns on the basis of these height estimates. A vertical gyroscope's pitch and roll measurements are, ideally, relative to the Earth's gravitational field. This alignment is not desirable in some circumstances, for example when the vehicle is experiencing a long-term change in terrain orientation such as climbing a long hill. It should however be possible to recursively estimate average values for the pitch and roll angles that reflect the long-term orientation of the surrounding terrain, using, for example,

$$\phi_{terrain} = 0.97\phi_{terrain} + 0.03\phi_{raw}$$
$$\psi_{terrain} = 0.97\psi_{terrain} + 0.03\psi_{raw}$$

Then the pitch and roll of the ith LIDAR are determined from the terrain orientation using

$$\phi_i = \phi_{raw} - \phi_{terrain} + \phi_{ioffset}$$
$$\psi_i = \psi_{raw} - \psi_{terrain} + \psi_{ioffset}$$

where $(\phi_{ioffset}, \psi_{ioffset})$ is the pitch and roll of the ith LIDAR with respect to the vehicle.

A normal vector in world coordinates for each horizontal LIDAR can be found by using the gyroscope measurements and the yaw measurement from the GPS/INS fusion algorithm as

$$N_{i=(1...3)} = \begin{bmatrix} \cos(\theta_i)\sin(\phi_i) - \sin(\theta_i)\cos(\phi_i)\sin(\psi_i) \\ \sin(\theta_i)\sin(\phi_i) + \cos(\theta_i)\cos(\phi_i)\sin(\psi_i) \\ \cos(\psi_i)\cos(\theta_i) \end{bmatrix}$$

where $(\phi_i, \theta_i, \psi_i)$ are the pitch, yaw, and roll angles of the ith LIDAR.

The normal vectors $N_{i = (1...3)}$ are used to define a plane for each LIDAR such that an approximation of the height for each measurement could be found by solving the following system of geometric plane equations (where $\bullet$ indicates the vector dot product)

$$N_{i=(1...3)} \bullet \begin{bmatrix} x_{ilidar} \\ y_{ilidar} \\ z_{ilidar} \end{bmatrix} = d, \, N_{i=(1...3)} \bullet \begin{bmatrix} x_{meas} \\ y_{meas} \\ h \end{bmatrix} = d$$

for height h, where $(x_{ilidar}, y_{ilidar}, z_{ilidar})$ is the location of the ith LIDAR and (x_{meas}, y_{meas}) is the measurement location. The resulting h is an approximate height because in actuality the (x_{meas}, y_{meas}) will be modified by the pitch and roll of the vehicle.

However, in actual experiments the estimated height measurement did not prove reliable. Errors in the pitch and roll angular measurements caused the object height error to increase linearly with range. This may have been due to the fact that the pitch and roll measurements were not synchronized with the sensor measurements, and the fact that the Crossbow gyroscope was in a shock-mounted enclosure that was not fixed rigidly with respect to the vehicle.

Another approach, which was deployed in the challenge vehicle, used a LIDAR scanning vertically to find a ground profile, which can be used to cull measurements that are close to the ground. Figure 4.30 shows a simulated vehicle with a vertically mounted scanning LIDAR pointing towards a small hill with a box structure placed in front of the hill. The raw output of the simulated LIDAR is shown in Figure 4.30(b). One can see that a scan of the vertical LIDAR finds a series of measurements that represent points on the ground. Since each LIDAR scan occurs at rate of 30 Hz, synchronization with the ground profile LIDAR measurements was not a major problem.

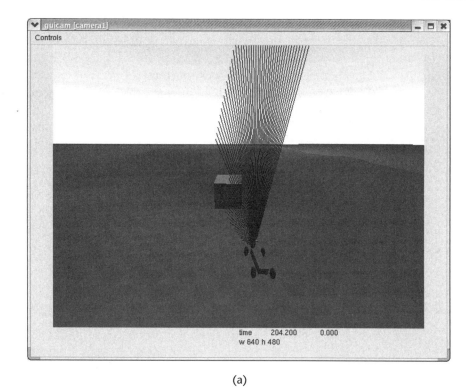

(a)

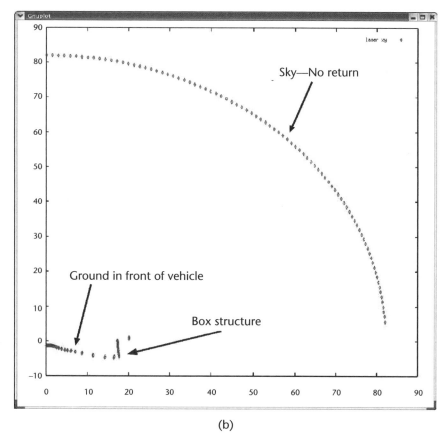

(b)

Figure 4.30   (a, b) Simulated vertically scanning LIDAR.

A windowed average filter is applied to the ground profile measurements to identify a large change in profile slope, which could signal an object or obstacle that should not be part of the ground terrain profile. Figure 4.30(b) shows a large change in slope where the LIDAR scan encounters the box structure in Figure 4.30(a). If a high slope object is detected, as in this case, the ground profile is linearly extended through the object using a slope that was established from the front of the vehicle to the point on the ground immediately preceding the object. It is critical that the profile of actual obstacles not be included in the ground profile, otherwise any measurements from that object will be discarded.

Further, the use of the ground profile derived from a single vertical LIDAR assumes that extending the ground profile laterally will provide a reasonable approximation to the terrain in front of the vehicle. This assumption breaks down whenever the sampled profile cross-section varies. However, this technique can be extended if multiple vertical LIDAR are available or if sufficient horizontal planes, as may be found on recent sensors such as the Velodyne HDL series, are available.

4.5.3.3   Sensor Fusion and Understanding

A block diagram of a sensor fusion system for an off-road application is shown in Figure 4.31. In this particular example, the result of the sensor fusion module was a cell-based internal map, 160 meters on a side, composed of 640,000 cells, each 20 cm on a side. Each cell contained three classes of confidence data: cell occupied confidence, C_o; cell empty confidence, C_e; and cell unknown confidence, C_u; as well as the estimated height, H_{object}, for an object within that cell. Placing sensor data into this map consisted of updating the three confidence measures, increasing or decreasing them as appropriate, and the obstacle height estimate.

4.5.3.4   Sensor Data Processing

The first step in sensor processing is to validate that the sensor is functioning and that the integrity of the data and communications between the sensor and processing equipment is intact. The second step is to remove any suspect or erroneous information. For the laser rangefinders, this processing includes the removal of ground clutter by culling points when

$$H_{object} < H_{terrain-profile} + 0.25\text{m}$$

that is, whose estimated heights are near or below the estimated ground profile. A stereo vision system, which produces a height above ground map, is similarly processed and areas that are marked as indeterminate are removed from the data stream. Most radar sensors deliver heavily processed results that do not contain ground clutter. A series of carefully constructed coordinate transformations then places all the sensor data into a unified frame of reference while helping to insure that the fusion results from multiple sensors do not distort or smear an object.

The fused LIDAR and vision data can be used both to detect objects and detect empty space. When a LIDAR or radar sensor detects an object, a square region of map cells of size

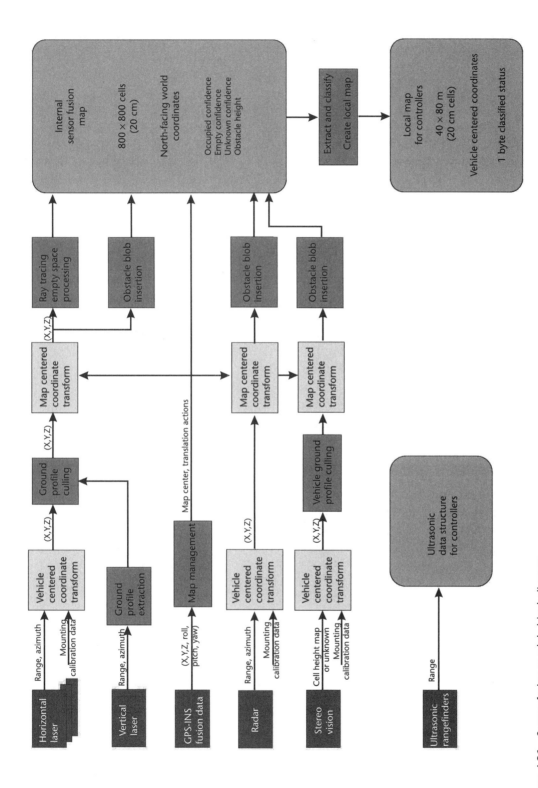

Figure 4.31   Sensor fusion module block diagram.

$$\frac{Range(meters)*\dfrac{\pi}{360}}{0.20}$$

is identified, representing the space in the map potentially occupied by the detected object. Data from the stereo vision system is already quantized into spacial rectangles.

For each cell in which the presence of an object is to be marked, the confidence values can be initialized as follows

$$C_e = 0.001, \quad C_u = 0.001, \quad C_o = 1.0$$

In order to properly handle empty space detection and avoid eliminating previously detected objects, the height of previously detected objects as well as the height of the scanning plane must be considered. The height of each cell is updated only when the height of the LIDAR scan is greater than the current height recorded in the map cell.

To process areas that are sensed to be free of objects, an efficient ray-tracing algorithm must be utilized such as Bresenham's line algorithm [18]. Some conditions must be defined to determine when the values for a cell that is sensed as empty are to be updated. For example, the rule for the OSU ION vehicle was chosen as

$$H_{object} < 2.0\,\text{m}$$
$$H_{object} > 0.0\,\text{m}$$
$$H_{object} < H_{cell} + 0.38\,\text{m}$$

When the test is successful, the confidence and height values are updates according to

$$C_o = 0.9C_o, C_u = 0.001 \ \text{ if } C_o > 0.001$$
$$C_e = 1.1C_e \ \text{ if } \ C_e < 1.0$$
$$H_{cell} = H_{object} \ \text{ if } H_{object} > 0.1\,\text{m}$$

Finally, temporal erasing (forgetting) is applied to the entire map according to

$$C_o = 0.998C_o \ \text{ if } C_o > 0.0001$$
$$C_e = 1.002C_e \ \text{ if } C_e < 1.0$$

These particular choices produce a sensor fusion output that is conservative with respect to removing already sensed objects from the world map.

4.5.3.5   Issues in Map Representation

A map representation that is convenient for sensor fusion may not be appropriate for local path planning and control functionality. For control and local path planning, a map representation with a vehicle-centered, body-fixed coordinate system where the location and orientation of the vehicle are fixed with respect to each map cell's coordinates may be most convenient. In general, the control algorithm is not terribly interested in the entire sensor fusion state, but rather simply whether a given region of space is traversable. Thus, it is useful to derive a temporary simplified map to be provided to the control software. This simplified map also requires less memory or communication bandwidth to transfer from one processor or software module to another.

However, while this coordinate system may be desirable for control purposes, it is not desirable for sensor fusion mapping. A body-fixed coordinate system requires that the raster map be rotated and translated each sample period as the vehicle moves. Rotated map cells, which do not exactly correspond to a set of integer coordinates, must be interpolated. The interpolation can be a source of error that builds over time as the vehicle moves through the world.

One solution to this problem is to keep the map in world coordinates. World coordinates have the advantage that no rotations or interpolations are required. Unfortunately, as the vehicle moves through the world, large numbers of additional map cells are revealed to the sensors. Since the map must occupy a fixed amount of memory, the vehicle must be periodically translated back to the center of the map. Thus, the map coordinates can be shifted from time to time such that the vehicle is kept approximately at the center of the map while the rotation of the map coordinates are kept fixed and aligned with world coordinates. The map coordinate translation shifts are integral shifts only and thus do not require any interpolation.

One rotation and translation operation would then be used to deliver the map information to the path planning and control algorithms. This operation is given by

$$X_{pos} = Rnd\left(\cos(\theta)(i - X_{mid}) - \sin(\theta)(H_{map} - j - Y_{mid}) + X_v\right)$$
$$Y_{pos} = Rnd\left(\sin(\theta)(i - X_{mid}) + \cos(\theta)(H_{map} - j - Y_{mid}) + Y_v\right)$$

where (i, j) are locations in the vehicle-centered, body-fixed output map, (X_{pos}, Y_{pos}) are locations in the approximately vehicle-centered, world-oriented internal map, θ is the vehicle yaw, (X_{mid}, Y_{mid}) indicate the location of the vehicle in the output map, and (X_v, Y_v) indicate the vehicle's location in the internal map. In this case, the interpolation is accomplished by rounding the result to the nearest integer.

While this operation does demand an interpolation, these interpolation errors do not accumulate because the results of this interpolation are not fed back to the sensor fusion map.

4.5.3.6   Sample Results

Some sample results are shown in Figure 4.32. They show the resulting sensor map on the top left, the raw sensor data from the LIDARs and the vision system on the

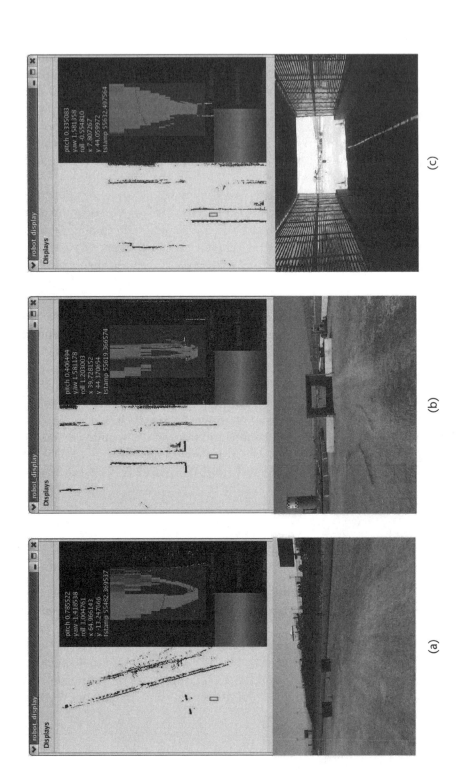

(a)

(b)

(c)

Figure 4.32 Sensor fusion results: (a) approaching a gate obstacle, (b) entering a tunnel, and (c) inside the tunnel (GPS blocked).

top right, and below that a visual image of the scene for comparison, for three examples selected from a 2005 DARPA Challenge qualification run at the California Speedway. The location of the vehicle on the map is indicated by a small horizontally centered box. The leftmost example shows the vehicle approaching a gate; the gate and the traffic cones are visible in the map, as well as two chain link fences separated by a walkway on the far right. Almost horizontal ground returns can be seen in the LIDAR data. Notice also that the gate creates an unknown shadow region in the vision system data. The middle example shows the vehicle approaching a tunnel. The tunnel, as well as jersey barriers behind it, can be seen in the map, along with fences and machinery along the right side. The rightmost example shows the vehicle inside the tunnel. GPS reception is unavailable, yet objects can be successfully placed on the sensor map. This example also clearly shows the memory capability of the sensor map, with objects behind the vehicle and occluded by the tunnel remaining in the sensor map.

4.5.4   Cluster Tracking and an On-Road Urban Vehicle

For an on-road vehicle, there is less variability in the target types that need to be detected and registered, and in general there is more a priori information available. However, the more structured nature of the environment implies that higher-level information must be extracted from the sensor data in order to accomplish the behavior and control objectives of the autonomous vehicle. In the on-road case, it is thus convenient to distinguish between the sensor fusion tasks (fusion, clustering, target classification, and target tracking algorithms), which evaluate and fuse the individual sensor data and create a unified representation, in this case a list of tracked clusters, and the situational awareness tasks (road and intersection model generation, track evaluation and classification, and condition evaluation and event generation) that produce information relevant to the control and behavior of the vehicle in its current situation.

4.5.4.1   Sensor Suite for an Example On-Road Vehicle: The OSU ACT

Ohio State University participated in the 2007 DARPA Urban Challenge, an urban autonomous vehicle competition held at the former George Air Force Base in California. The overall behavior of the vehicle deployed in the competition is generated using a nested finite state machine architecture implemented in the high-level control module as described in Chapter 3. Decisions and transitions in this system are driven by trigger events and conditions extracted by the situation analysis module from a model of the external world build from an array of sensors by the sensor fusion module, as well as conditions extracted by the high-level control module from the vehicle's position, mission, and map databases. An overall block diagram of the vehicle is shown in Figure 4.33.

The on-vehicle sensors must provide both information about the state of the ego vehicle, including its position, velocity, and orientation, and coverage of all features of interest in the external world. This information must be extracted and provided to controllers in real time.

The current position, orientation, and velocity of the vehicle is estimated using a Kalman filter–based software module analyzing data from a number of sensors.

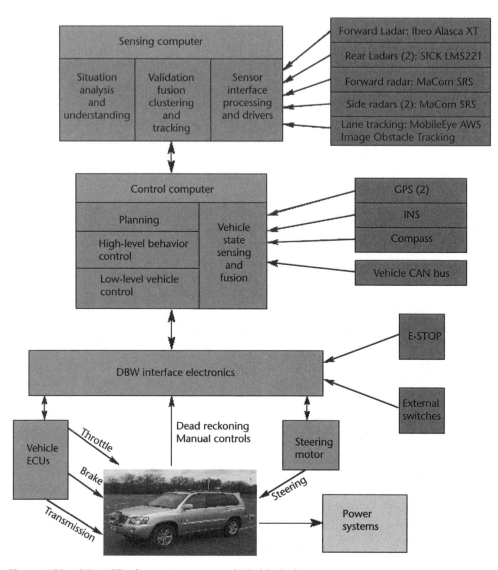

Figure 4.33   OSU ACT urban autonomous vehicle block diagram.

Two Novatel Propak-LB-L1L2 GPS receivers using Omnistar HP differential correction technology provide direct redundant position and velocity measurements. A Crossbow VG700A three-axis fiber-optic vertical gyroscope provides angular rate and linear acceleration measurements as well as internally estimated roll and pitch angles in an SAE Earth-fixed coordinate system. Independent wheel speed measurements for each wheel, along with overall vehicle velocity, transmission, and steering wheel angle are also obtained from the vehicle CAN bus. These measurements may be augmented with measurements from the vehicle's own stability system gyroscopes and accelerometers, which are also obtained from the vehicle CAN bus, although the noise and bias of these sensors are significantly larger than those of the high-grade GPS and IMU units. Sensor validation software is used to eliminate sensor failures and errors, especially GPS related step-change events caused by changes in differential correction status or the configuration of the visible satellite constellation.

The OSU-ACT vehicle is equipped with a sensor system that attempts to completely cover the area around the vehicle, as shown in Figures 4.34 and 4.35, and to meet, within budgetary constraints, the requirements for vehicle behavior control. A forward looking scanning laser rangefinder, the Ibeo Alasca XT, provides 4 vertical scanning planes, a range of up to 160 meters, an azimuth resolution under 0.25°, and a horizontal field of view (as mounted in front of the grill) of 220°. Two SICK LMS 221-30206 scanning laser rangefinders are mounted behind the rear bumper and angled at 20° from the lateral axis of the vehicle to provide rear and side sensing with a range up to approximately 60 meters and an azimuth resolution of 0.5°. Two side-mounted MaCom short-range sensor radar systems, with a range of approximately 25 meters and a horizontal field of view up to 70°, provide side sensing of vehicles in adjacent lanes and help to fill the gaps in coverage of the LIDAR sensors close to the vehicle. A forward mounted MaCom short-range sensor radar sensor provides redundant coverage for the Ibeo LIDAR. In addition, a Mobileye AWS image processing sensor system is installed to both detect forward vehicles and to provide lane marker extraction. OSU developed lane marker tracking may also be installed if needed, as well as image processing algorithms developed by other team partners.

Figure 4.36 shows an example of sensor returns from all three LIDARs combined into a single vehicle centered display. The solid rectangle in the middle of the display is the ego vehicle. The front of the vehicle is towards the top of the display.

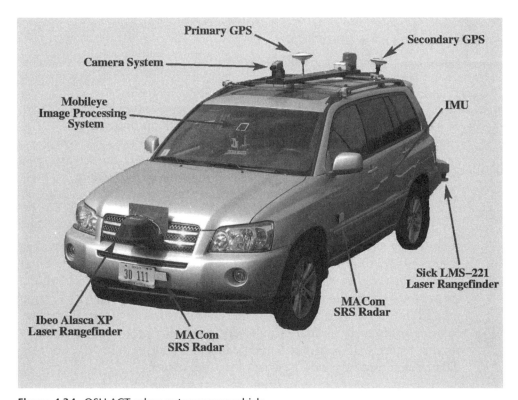

Figure 4.34 OSU ACT urban autonomous vehicle.

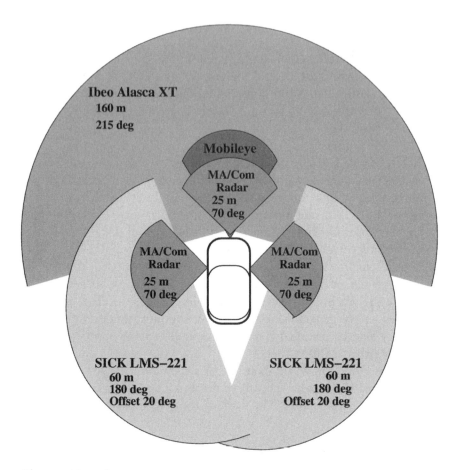

Figure not to scale.

Figure 4.35   Sensor suite and sensor footprints.

Figure 4.36(a), acquired while the vehicle was in a parking lot, shows other vehicles as well as a building to the rear of the vehicle using outlines for the LIDAR returns. Figure 4.36(b), acquired at a multilane intersection, shows other vehicles, including summary information from the track of one vehicle target, as well as ground cluster from curbs, small hills, and vegetation on the sides of the roads.

4.5.4.2   Sensor Fusion and Object Tracking

A clustering and tracking approach is useful in traffic situations when the environment is urban, road-network based, and highly dynamic. The sensor fusion module in this case is responsible for clustering and tracking all objects that are seen by the sensors. When the primary sensors are scanning laser range finders positioned around the periphery of the vehicle, the clustering must be done in the fusion algorithm. Sensors such as radar systems tend to already produce clustered and tracked results. Figure 4.37 shows a general data flow and the transformations as sensor data moves from raw form to a list of tracked objects.

The first step uses the position and orientation estimates from the fusion of GPS and INS information to place the sensor measurements into a coordinate

(a)

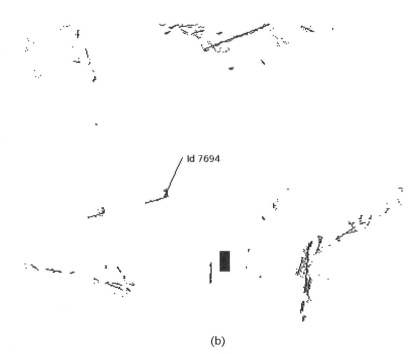

(b)

Figure 4.36   Sensor returns (a) in a parking lot and (b) at an intersection.

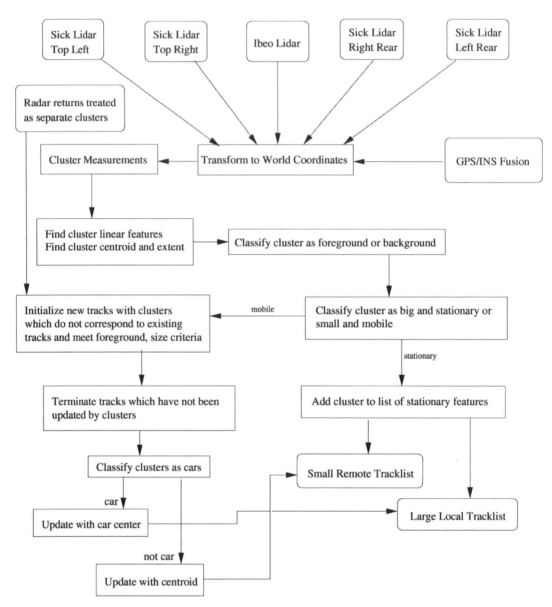

Figure 4.37   Sensor fusion and tracking block diagram.

system that is fixed with respect to the world. Once the sensor measurements are in a world-fixed coordinate framework, the sensor measurements can be placed into clusters that form a group of returns (within 90 centimeters of one another in the case of the algorithm deployed on the ACT vehicle). Various spatial descriptors can then be extracted from each cluster. Linear features can, for example, be extracted with a RANSAC-based algorithm. The geometry and extent of the cluster can be recorded, the cluster centroid calculated, and the linear features used to estimate a cluster center if the cluster is determined to represent another vehicle. Large clusters, of a size that is unreasonable for a road vehicle, should not be tracked as vehicles. Instead the large clusters can be reported as objects with zero velocity and the linear features of the large clusters can be placed into a "clutter line" list that can terminate the tracks of any vehicles that venture too close.

After the sensor returns have been clustered, each cluster is classified according to whether it is occluded by another cluster. Clusters that are not occluded by another cluster are classified as "foreground" while clusters that are occluded on the right are "right occluded" and on the left are "left occluded."

The principle output of the sensor fusion algorithm is a list of tracks. Each of the resulting tracks has a position and velocity. Also, the general size and shape of the point cluster supporting the track is abstracted as a list of linear features. One implementation strategy for each of these tasks is described in the rest of this section.

4.5.4.3   Transform to World Coordinates

The position and orientation of the vehicle can be used to place measurements into world coordinates. The first step is to move the sensor measurements from the local sensor coordinate frame to a vehicle-centered frame. The actual transform depends on the conventions used by the sensor and its location on the vehicle. As an example, the sensor to vehicle coordinate transform for a SICK scanning LIDAR is:

$$X_{vehicle} = r\cos(\theta + \theta_{offset}) + X_{offset}$$
$$Y_{vehicle} = -r\sin(\theta + \theta_{offset}) + Y_{offset}$$

where (θ, r) are sensor hits returned by the LIDAR and $\theta_{offset} = -116$, $X_{offset} = -1.93$, $Y_{offset} = -0.82$ are the location and orientation values for the LIDAR, in this example the left rear SICK LIDAR on the ACT vehicle. These offset values can be obtained with physical measurements as well as experimental calibration tests involving the detection of objects in the overlapping regions of multiple sensors. For instance, if one is confident about the orientation offset of one sensor, the offsets of another sensor can be discovered by observing the same distinctive object within some overlapping region.

Other sensor-to-vehicle coordinate transformations are similar to the transform described above, but they are specific to a certain sensor. A sensor's range measurement needs to be scaled to meters, and the angle measurement's direction may be reversed, or there may be an angle offset. One choice for the vehicle-based coordinate system is the standard SAE coordinate system shown in Figure 4.38. The positive x-axis passes through the front of the vehicle, and the positive y-axis exits the right side. The positive z-axis points down and the yaw axis is positive clockwise.

The transforms for moving to world coordinates from vehicle coordinates are the following:

$$X_{world} = -X_{vehicle}\sin(\phi) + Y_{vehicle}\cos(\phi) + X_{position}$$
$$Y_{world} = X_{vehicle}\cos(\phi) + Y_{vehicle}\sin(\phi) + Y_{position}$$

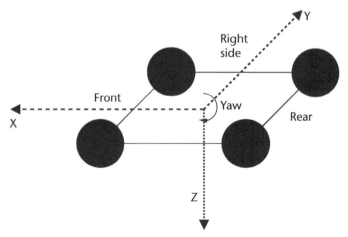

Figure 4.38 SAE vehicle coordinate system.

where $(X_{position}, Y_{position}, \phi)$ are the position and orientation of the vehicle as returned by the GPS/INS fusion algorithm.

4.5.4.4   Clustering LIDAR Returns

It is possible for there to be many thousands of measurement points returned by the LIDARs in a given scan period. In order to effectively track the targets, these points need to be formed into clusters. A point can be considered to belong to a cluster if the point is within a certain distance of a point already in the cluster. The parameter value represents a compromise between the possibilities of accidentally merging two adjacent vehicles together versus the danger of dividing a vehicle into multiple clusters due to partially missing sensor returns. For example, during the development of the ACT vehicle a value of 90 cm was found to produce the best results over a significant amount of collected urban traffic data. An algorithm that can be used to cluster the LIDAR measurements is described in the pseudo code below. The code relies heavily on the Union-Find [19] algorithm to place LIDAR returns and maintain information about the disjoint sets that represent clusters.

```
place sensor measurements in vehicle coordinates in polar form
sort by angle
   for each point p_i
     Set_1 =FIND(p_i)
     for each point p_j in the vicinity of p_i
        Set_2 =FIND(p_j)
        if set_1 is not equal to set_2 and
       if the difference in range and angle between p_i and p_j is small
         enough
           UNION(set_1,set_2)
           Set_1 =FIND(p_i)
```

In the case of sensor measurements, the above code performs reasonably well while still minimizing computational time by prioritizing angular comparison. Points that are adjacent in the angle axis of their polar representation are more likely to be in the same cluster than points that are not. This property allows the algorithm to limit the comparisons between p_i and p_j to only those returns for which p_j is in the angular vicinity of p_i. In order to achieve computational efficiencies, the notion of a neighborhood can be defined in terms of the index into a list of points sorted by angle.

However, the above pseudo code doesn't guarantee a correct clustering. In pathological cases it is possible for multiple clusters to be derived from points that should be within a single cluster. Thus, a correction step can be performed in which all the clusters found with the above pseudo-code are checked to see if they are closer than the threshold distance parameter value defined above. Clusters identified in this process can be merged. Since the number of clusters will be fairly small in comparison to the number of sensor measurements, the correction step will not be terribly time consuming. During experiments on the ACT vehicle, this correction was rarely required in practice.

4.5.4.5 LIDAR Cluster Interpretation

The sensor measurements do not completely describe the shape of a target vehicle. The scanning laser rangefinders, which provide the most detailed image of an object, cannot see points on the far side of a target object—that is to say, the sensor image of an object does not show the complete shape of the object. Therefore, it is not always easy to identify an appropriate point to track within a cluster of points. The best views of a target vehicle are often L-shaped as shown in Figure 4.39.

Figure 4.39 also shows that the most obvious choice for a point to be tracked, the average location of points in the cluster, the centroid, moves within the target vehicle as the target vehicle displays different orientations to the ego vehicle. There are other cases in which the centroid of a cluster may move independently of the object responsible for the cluster [20], including when an object is partially

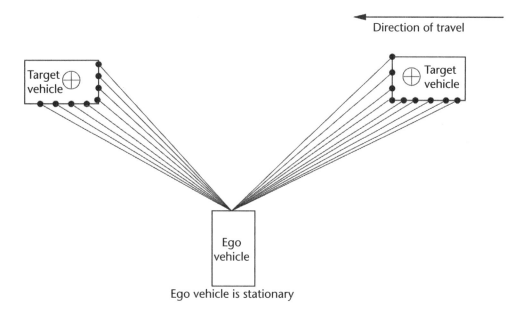

Figure 4.39 Observation angle versus cluster centroid.

occluded or is moving into or out of occlusion (part of a closer cluster "shadows" a more distant cluster), and when an object is large enough to extend past the range of the LIDAR (for example a hedge or barrier on the side of a road).

Thus, the computed centroid of a cluster does not provide a reasonable tracking point for a tracking filter because the centroid is moving with respect to the underlying vehicle and confounding the vehicle's estimated position and velocities with non-Gaussian noise. The fact that changes in the target orientation change the shape of the target makes it necessary to fill in unseen parts of the target shape with estimates.

For this particular application, given the target-vehicle-rich environment, it is reasonable to define a fairly simple test to classify clusters as vehicles. One possible test is to find linear segments and decide, based on linear segment length, whether the linear segment represents some facet of an automobile. For example, in the algorithm deployed on the ACT vehicle, if the tracker sees a cluster that has a line segment greater than 1.25 meters and less than 7 meters, the tracker classifies the track as a car. Vehicles or objects that possess linear features larger than 7 meters are tracked, but they are not classified as cars.

The random-sample-consensus (RANSAC) approach is a good choice for finding line segments in a cluster. Below is pseudo-code for finding the most prominent linear feature in the point cloud data for a cluster:

```
For a set number of RANSAC iterations
  line=find a hypothesis line from the unused points supplied to this
  function
  number_of_support_points=evaluate the line
  if number_of_support_points>max_number_of_support_points
    max_number_of_support_points=number_of_support_points
    best_line=line
return best_line
```

After the above code returns with a prominent line, the support points that are within some minimum distance of this line can be flagged as "used" and the above function called iteratively until the number of support points drops below some threshold. After a set of lines is built up in this way, each of the lines may be broken into at least one line segment.

As mentioned above, the obvious choice for the cluster's position, the cluster centroid, presents problems for a tracking filter. Nevertheless, in some cases it may be the only available measure. We give one possible approach to handle this problem.

If no suitable linear features are found within the cluster, the tracker attempts to track the cluster centroid that is given by

$$X_{centroid} = \frac{1}{n}\sum x_i$$

$$Y_{centroid} = \frac{1}{n}\sum y_i$$

If a small (<80 cm) linear feature is found in the cluster, such as might be a return from part of the front or back side of vehicle, a center point is found by extending a line in the direction away from the ego vehicle for a distance of 2 meters, as shown in Figure 4.40(a). This type of center is called the *line midpoint*. The assumption behind the line midpoint center is that while the cluster is probably a vehicle the small linear feature representing only a small part of the car presents an unreliable orientation. If a linear feature is found that is long enough to represent the side of a car, the *Car Side* center case derives the vehicle's center location from the linear feature as shown in Figure 4.40(b). The car side center is located 1 meter from the midpoint of the linear feature. Similarly, if a linear feature is found that is long enough to represent a significant portion of the front or back side of a vehicle,

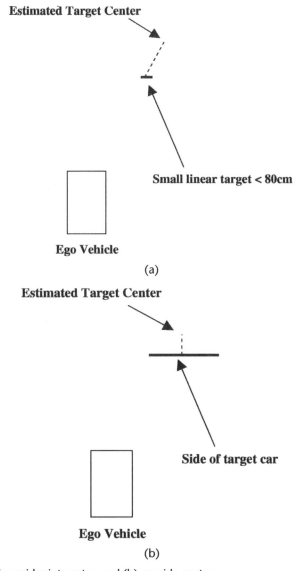

Figure 4.40   (a) Line midpoint center, and (b) car side center.

the *car end* center point case identifies the center point as located 2 meters from the midpoint of the linear feature. Once a track has been classified as a vehicle, its classification can be fixed regardless of later changes in observed features.

The target vehicle's orientation can be estimated by analyzing the direction of the track's velocity vector and the direction of the cluster's linear features. The track's velocity vector has the advantage of being accurate when the target is moving. However, when the vehicle is not moving, the orientation estimate must rely on the direction of the cluster's longer linear features. The cluster's linear features may represent the sides or ends of a vehicle. However, with a prior estimate for the vehicle orientation the forward/backward ambiguities of the linear features may be resolved. Pseudo-code for finding an orientation estimate from the geometry of the linear features is:

```
proc vector direction_from_geometry(current_estimated_direction, best_linear_feature)
    if the best_linear_feature exists and is longer than 0.5 meter
        if best_linear_feature length is longer than 2.25 meters
            assume that the best_linear_feature represents the side of the car
            find geometry_direction
        else
            assume that the best_linear_feature represents the end of the car
            find geometry_direction
        if the dot product of geometry_direction and current_estimated_direction < 0
            flip the direction of geometry_direction 180 degrees
    else
        geometry_direction=current_estimated_direction
return geometry_direction
```

Once a direction estimate has been extracted from the linear features, this linear feature direction estimate may be used to update the final orientation estimate in situations where the direction estimate obtained by looking at the target's velocity is poor. Specifically, the orientation can be taken to be the direction of the velocity estimate whenever the speed is above a chosen speed (2 meters per second in the deployed ACT algorithm) while the geometry-based direction is used at slower speeds.

4.5.4.6 Tracking with the Clusters

Once the sensor measurements have been clustered, each measurement cluster should represent a single target of interest and the resulting targets must be tracked. The target clusters can then be matched to the list of existing tracks. Clusters that cannot be matched to an existing track are used to initialize new tracks. This one-to-one cluster to target assumption is not strictly true. It is possible for a single target to be divided into two clusters if a closer object is occluding it. However, the assumption is true often enough to move forward with a cluster tracking system.

The process of drawing correspondences between measurement clusters and existing tracks is known as *data association*. One approach to this problem is based on the linear assignment problem. In this approach, we need to find a column mapping that minimizes the diagonal of a cost matrix A where element a_{ij} is the cost of associating cluster i with track j. The column mapping solution can be found using the linear assignment algorithm described in [21]. A simple choice for the cost of associating cluster i with track j is the distance between the centroid of the cluster i and track j. However, an ad hoc association cost that was used in the ACT deployed algorithm was:

$$a_{ij} = 100C_{dist} + \frac{100}{N_p} + 60|S_c - S_t|, \; all \; N_p > 0$$

$$a_{ij} = 100C_{dist} + 100 + 60|S_c - S_t|, \; all \; N_p = 0$$

where C_{dist} is the distance between cluster i and track j in meters, N_p is the number of points in the track j, S_c is the extent of cluster i in meters, and S_t is the extent of the last cluster that updated track j in meters. The motivation for the $100/N_p$ term is to make associating an old track with a cluster easier than associating a new track. The motivation of the $|S_c - S_t|$ term is to make it easier to associate tracks and clusters of similar sizes. Note that, generally, there are not equal numbers of clusters and tracks, so the matrix A is padded with entries of maximum cost value such that it is square.

The tracking filter could involve instantiating a complete Kalman filter for each track. However, a computationally simpler approach is to use a simple variable gain Benedict-Bordner [22] filter, also known as an alpha-beta filter, to estimate the current states of the track. These current track states are then used to match tracks to the current set of measurement clusters. The coefficients in the filter's update equations can be changed heuristically according to the changing classification status of the cluster.

The Benedict-Bordner track update equations are:

$$\hat{v}_x = v_x + h(p_{x\_meas} - p_x)/T$$
$$\hat{v}_y = v_y + h(p_{y\_meas} - p_y)/T$$
$$\hat{p}_x = p_x + Tv_x + g(p_{x\_meas} - p_x)$$
$$\hat{p}_y = p_y + Tv_y + g(p_{y\_meas} - p_y)$$

where (v_x, v_y, p_x, p_y) are the current velocity and position of the target, and $(\hat{v}_x, \hat{v}_y, \hat{p}_x, \hat{p}_y)$ is the updated velocity and position of the target, T is the sample period, and (p_{x_meas}, p_{y_meas}) is the centroid of the updating cluster. The values g and h are filter parameters, which can be heuristically modified during the evolution of the track.

The point within the cluster that is being tracked can change as the track evolves. However, all of the various types of tracking points described above attempt to approximate the center of the target. The least desirable tracking point is the cluster centroid, while more accurate tracking points attempt to use a line segment representing the side or end of a vehicle. The g and h parameters control the degree to which the filter will rely on internal states when updating. For the algorithm deployed in ACT, the parameters for various target center classifications are shown in Table 4.3. Note that if a vehicle has recently changed states, it may be prudent to not update the velocity estimate at that time (essentially this is $h = 0$). This allows the large artificial velocity that would be generated during center type transitions to be ignored.

4.5.4.7   Clutter Lines

Most returns from the vehicle's environment will be from untrackable objects such as vegetation, embankments, buildings, and so forth. Clusters generated by these returns are generally known as clutter. Any cluster that contains a linear feature that is longer than a reasonable length for a road vehicle, for example, 10 meters, should be assumed to be clutter and neither matched to an existing track or used to initialize a new track. However, these large objects are significant returns, which could be of interest to control algorithms, so these large objects can be reported as clusters with zero velocity.

The clutter linear features can be collected as a list of "clutter lines." The clutter lines are features that can terminate any target tracks that venture too near the clutter. In our experience it is unlikely that a target track will successfully transit the clutter without the target state being hopelessly contaminated by the nontarget sensor measurements. Terminating such a track may be considered a drastic step, but in our experience is preferable to allowing the existence of wildly erratic tracks. Examples of clutter lines which represent large objects, for example, road curbs and ground rising away from the road bed are seen in Figure 4.42, which was collected as the experimental vehicle approached a large intersection.

4.5.4.8   Occlusions and Background/Foreground Classification

Tracks occlude one another frequently. While tracks can be propagated with estimated positions for a small period of time during occlusions (2 seconds in the ACT deployed algorithm), it is likely that these tracks will acquire an incorrect velocity during the process of moving into occlusion. Figure 4.41 illustrates the problem. As vehicle 2 passes behind vehicle 1, the calculated center slows while the target does not. Thus, it is necessary to find the occlusion status of each cluster. The occlusion status can be found with an angular z-buffer where the buffer's bins represent angles and the value of the bin is the range.

The pseudo-code for creating such a z-buffer is shown here:

Table 4.3   Filter Parameters Versus Track Type

Center Type	g	h
Centroid	0.15	0.025
Line midpoint	0.2	0.05
Car side	0.3	0.1
Car end	0.3	0.1
Not a car	0.3	0.01

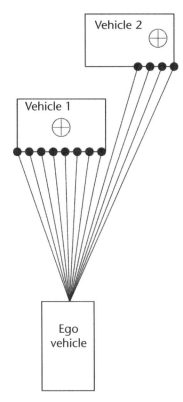

Figure 4.41   Example of occlusions.

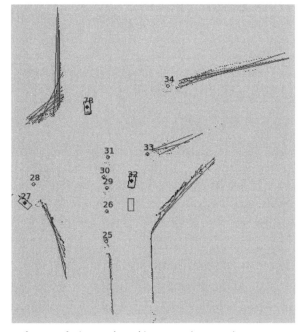

Figure 4.42   Output of sensor fusion and tracking at an intersection.

```
proc add_point_to_z_buffer(angle_val, dist_val)
find a bin for angle_val
assign this bin to variable i
if zbuffer[i]>dist_val then assign zbuffer[i] to be dist
```

All the bins in "zbuffer" are initialized to very large values prior to any calls to "add_point_to_z_buffer." After the above code has been called for all returned LIDAR points, "zbuffer" can be used to find the occlusion status of the left and right sides for each cluster with the following routine described as pseudo-code:

```
proc foreground(current_cluster)
find current_cluster's maximum angle as a zbuffer bin index
assign this bin index as i
find current_cluster's minimum angle as a zbuffer bin index
assign this bin index as j
check zbuffer from bin i to bin i+4 to see if any objects are closer,
   if so, mark as
      RIGHT_SIDE_OCCLUDED
check zbuffer from bin j to bin j-4 to see if any objects are closer,
   if so, mark as
      LEFT_SIDE_OCCLUDED
```

The occlusion status of the cluster can then used to inform track initialization decisions and modify the values of the g, h parameters for the tracking filter. For instance, pseudo-code for the "trackable_object" method used to decide whether to initialize a track for a cluster follows:

```
boolen proc trackable_object
if the cluster is large clutter return false
if the extent of the cluster is large return false
if the cluster has less than 2 points return false
if the cluster is foreground only return true
if the cluster is left occluded with greater than 10 points return
   true
if the cluster is right occluded with greater than 10 points return
   true
if the previous conditions failed to occur return false
```

If a cluster is left or right occluded, the values of tracking filter parameters g and h can be reduced (for example, by 0.1 in the ACT deployed algorithm) to make the track's states less sensitive to measurement updates.

4.5.4.9 Sample Results and Summary

Figure 4.42 shows sensor measurements along with some interpretation annotations. The measurements are in vehicle-centered coordinates and the rectangle in the middle of Figure 4.42 is the ego vehicle. The small dots are LIDAR measurements. The locations that are indicated with black target ID numbers are the locations of active tracks. Target ID numbers that also have black rectangles associated with them are interpreted as vehicles because the cluster that updated the track possessed linear features of the appropriate length at some point in the track's evolution. For instance, Figure 4.42 shows a vehicle, track ID 32, stopped at an intersection in front of the ego vehicle.

4.6   Situational Awareness

The term *situation* is defined to be knowledge concerning the vehicle and/or the prevailing scenario and surroundings. On the cognitive level, the achievement of situation awareness is considered an important open issue by many researchers [23]. Vehicular sensing should not be restricted to conventional metrology that acquires a couple of parameters from the scene, such as position and speed of other vehicles. Instead, a transition to consistent scene understanding is desirable. This requires a consistent propagation of knowledge and measures for its confidence throughout the perception chain as depicted in Figure 4.43. An ambitious goal is to expressively formulate the ambiguity of knowledge at every level that might stem from sensor noise, the signal interpretation process, or the ambiguity of previous levels. Thus, safety measures can be assigned to any potential behavior at the control level considering all previous processing. In the reverse direction, selective information can be required from previous processing units (e.g., sensor attention may be directed towards the most relevant objects in the scene). The numerous closed loops in this structure may motivate concerns and require a theory on perception stability.

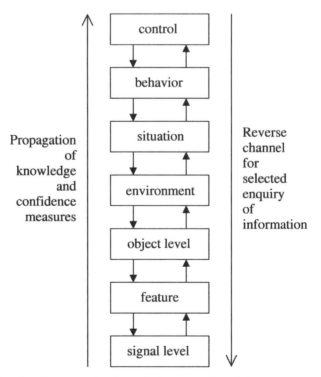

Figure 4.43   Bidirectional propagation through the perception chain.

For an off-road scenario, the situation is always path following with obstacle and collision avoidance. The behavior and control algorithms are required to analyze the occupancy grid map, identify obstacles that may block the current path, and adjust the planned path as needed.

For an on-road scenario, we are interested in all the targets in our path and the targets in surrounding lanes or on roads intersecting our lane. We are not interested in targets that do not affect the current situation and planned behavior. While an autonomous vehicle is navigating through an urban environment, many different situations may arise. The situations may vary if the vehicle is on a one-lane road, on a two-lane road, at an intersection, and so on. Particularly critical for an autonomous vehicle are those situations related to intersections. When a car is approaching an intersection, it must give precedence to other vehicles already stopped. If the intersection is not a four-way stop, the vehicle must cross or merge safely in the presence of oncoming traffic. If other vehicles are stationary for a long time, the car must decide whether those vehicles are showing indecisive behavior. Other situations may involve road blockage in which the vehicle might carefully perform a U-turn, park in parking spaces, and deal with dangerous behavior from other vehicles. All these situations must be identified and evaluated and the resulting conclusions transmitted to the high-level controller in order for the vehicle to operate properly.

From a practical viewpoint, situations and events are the switching conditions among metastates and all the substates inside the high-level behavior control state-machines. Thus, the aim of situation analysis is to provide the high-level controller with all the switching conditions and events in a timely manner. The situation analysis software must analyze the current vehicle state, the current and upcoming required behavior for the route plan, the map database, and the sensor data to identify specific situations and conditions that are relevant to the vehicle's immediate and planned behavior. In order to reduce computational costs and complexity, only the situations related to the current metastate or substates, as provided by the high-level control software, should be checked.

4.6.1 Structure of a Situation Analysis Module

The overall inputs and outputs for a situation analyzer system are illustrated in Figure 4.44. We may assume that the sensor fusion and tracking system provides information about the set of tracked objects. Information about position and velocity of the cluster, combined with the location of the lines defining the geometry of the cluster, are given in a local navigation reference frame. We may also assume that the path planning algorithm provides information related to the current optimal path. The path is a sequence of links defining a traversable set of roads. Starting from the path, the situation analyzer can identify the location of the road. A road model can be structured as a set of polygons generated from a spline curve fitting the waypoints in the optimal path. Such a road model design is particularly suitable for both accuracy and implementation purposes. Using such a road model, traffic vehicles and other obstacles can be located and classified according to their lane positions and their impact on the current behavior and plan for the autonomous vehicle's motion can be analyzed. Finally, we can assume that the high-level behavior control provides information about the current state of the vehicle. In order to keep

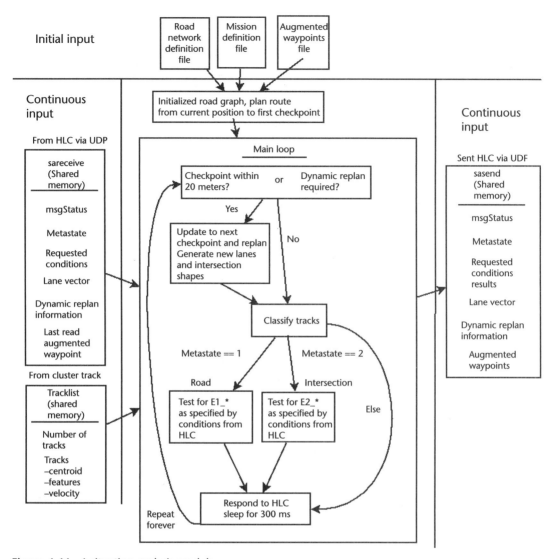

Figure 4.44   A situation analysis module.

the computational cost as low as possible, only the situations related to the current metastate or substates are checked.

Figure 4.44 shows the algorithm flow of a situation analysis (SA) software module. In the case of the algorithm deployed on the ACT vehicle, the high-level control is developed and implemented as a nested finite state machine with multiple metastates. The high-level control executes on a different processor. To determine the conditions for changing states within the metastates, the high-level control sends network UDP packets to the computer on which the sensor fusion and situational analysis algorithms are executing requesting certain conditions and providing information about the current state and the current metastate. As shown in Figure 4.44, the first activity is to generate lane and intersection models as the vehicle moves through the world. After this, requests from the high-level controller are received and the requested tests are executed. Finally, a network UDP packet is sent back to the high-level control algorithm containing a reply to the requested conditions.

In many cases the output of the situation analysis algorithm is a series of Boolean (true/false) response values. For example, if the vehicle is currently driving down a road and about to perform a passing operation, the current state might be "give turn signal and check passing lane." The situation analysis algorithm might then be asked to evaluate three conditions: is the passing lane temporarily occupied, is the passing lane permanently occupied, and is there sufficient space to pass before an intersection.

4.6.1.1   Comparison to Other Methods

The 2-D occupancy grid is used as a common approach for obstacle avoidance and situation awareness/analysis. The 2-D occupancy grid requires a matrix-like data structure that covers the entire map of a possible route. This data structure can have uniform size grid squares or variable size squares where a smaller grid would be used in an area of interest. An example of the occupancy grid is described in [24]. Using the occupancy grid with a defined road network can consume a significant amount of memory for areas that can never be reached by an on-road vehicle. Also, determining if a track is in any of those possible grid squares can be time consuming and may not be necessary for the current situation.

To avoid those problems, the approach outlined above focuses solely on the route that will be followed and expands the area to be evaluated only when the situation requires it. For example, in a normal lane following operation we will only construct the current roadway and look for obstacles in that area. If we approach an intersection, an obstacle is detected, or more generally the situation changes, a new area will be generated or loaded and evaluated.

4.6.2   Road and Lane Model Generation

One of the most important issues in the situation analyzer is road model generation so that target tracks may be classified and analyzed relative to the road lanes and driving requirements. A good road model, closely matching the actual road, is fundamental to safely driving in an urban area. The road can be generating starting from the optimal path, which is determined from the starting position to one or more goal locations using a path planning algorithm operating on a road map database. The map database is assumed to provide lane center waypoints, though these may be sparse in some situations, and lane widths. An example of this, taken from the DARPA Urban Challenge course, is shown in Figure 4.45, in which the lanes are overlaid on a satellite image of the area.

From the links in the optimal path, the list of the consecutive optimal waypoints can be extracted. For an accurate road generation we need close points, while the given waypoints may be very far from each other. The situation analysis module can generate a spline or linear fit, based on the sharpness of the curvature between each point, passing through each known waypoint. Catmull Rom splines [25–27] are suited to this task, as they are unique from other mathematically computed arcs in that they pass through all of their control points. Moreover the spline is C^1 continuous, meaning that there are no discontinuities in the tangent direction

Figure 4.45   GIS map with waypoints and actual path traveled.

and magnitude. However, the spline is not C^2 continuous. The second derivative is linearly interpolated within each segment causing the curvature to vary linearly over the length of the segment. This is illustrated in Figure 4.46.

To determine the sharpness of a turn, three waypoints can be used to create two vectors. The normalized dot product of these two vectors will give the cosine of the angle between these vectors. If the absolute value of the cosine is less than some value (in the deployed ACT algorithm 0.9, representing an angle of greater than 25°) a simple linear connection can be employed, otherwise a Catmull-Rom spline can be used to connect the waypoints. The equation for the Catmull-Rom

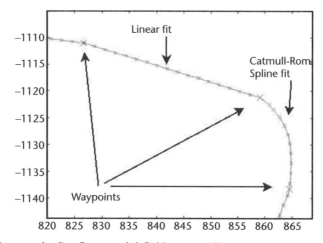

Figure 4.46   Linear and spline fit to road definition waypoints.

spline requires four control points, one waypoint before the beginning of the turn, one waypoint at the start of the curve, one waypoint the end of the curve, and one waypoint after that. Using these control points the equation for the x component of the spline is:

$$X_{sample} = \frac{1}{2}\left(\begin{array}{l} \left(-t + 2t^2 - t^3\right)Xwp_{i-1} + \left(2 - 5t^2 + 3t^3\right)Xwp_i \\ + \left(t + 4t^2 - 3t^3\right)Xwp_{i+1} + \left(-t^2 + t^3\right)Xwp_{i+2} \end{array} \right)$$

where t is a variable parameter. The variable t can be searched iteratively until a desired sample distance has been traversed, at which point another sample of the road location is taken. The sample distance used in the deployed ACT vehicle is 2 meters in order to generate a highly accurate road model. The spline function is thus sampled with equally spaced points. If a linear fit is performed in a particular region, the line can be sampled analogously.

There are a number of possible road model designs. One possible approach is to consider a road map with fixed memory occupancy. The map can be divided into geometric cells. Depending on the desired accuracy, the number of these cells may be very large. Since the number of the cells is fixed, the map must be very frequently updated during the mission. Moreover, in this approach there might be many empty cells in memory carrying no information about the road shape or occupancy.

A better choice, accurate and computationally simple, is to represent a road by a set of polygons. From each pair of consecutive samples, a rectangle is constructed. Two consecutive sample points (x_1, y_1) and (x_2, y_2) are used to calculate a normal (X_{left}, Y_{left}) from the line they create using:

$$Slope = \frac{y_1 - y_2}{x_1 - x_2}$$

$$B = y_1 - Slope * x_1$$

$$Offset = \frac{x_1 + x_2}{2Slope} + \frac{y_1 + y_2}{2}$$

$$\alpha = 1 + \frac{1}{Slope^2}$$

$$\beta = -2\frac{x_1 + x_2}{2} - 2\frac{Offset}{Slope} + 2\frac{y_1 + y_2}{Slope}$$

$$\gamma = \left(\frac{x_1 + x_2}{2}\right)^2 + Offset^2 + \left(\frac{y_1 + y_2}{2}\right)^2 - 2Offset\frac{y_1 + y_2}{2} - 4$$

$$X_{left} = \frac{-\beta + \left(\sqrt{\beta^2 - 4\alpha\gamma}\right)}{2\alpha}$$

$$Y_{left} = -\frac{1}{Slope}X_{left} + Offset$$

With the normal and the lane widths defined in the map database, four new points can be created orthogonal to the samples and half a lane width away. A sample result is shown in Figure 4.47. To compensate for the missed and overlapping areas shown in Figure 4.47, the intersection of the projections of each polygon's left side and right side can be calculated, thereby converting rectangles to trapezoids to eliminate overlapping and to fill missed areas. This result is shown in Figure 4.48. The individual polygons are called road stretches.

We also need to generate a model of any lanes that are adjacent to the current lane, as they could be used as a passing lane at some point. To generate these lanes, the closest waypoints next to the current lane waypoints are found and the lane is sampled in the same fashion as the current lane. To create the polygons, however, the current lanes stretches are extended and the intersection with the new lane is found. This allows one to link the current lane stretches to the adjacent lane

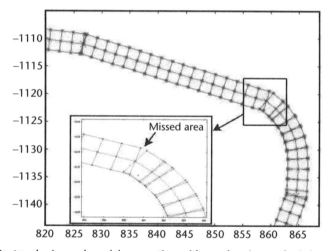

Figure 4.47   Rectangles in road model generation with overlapping and missing regions.

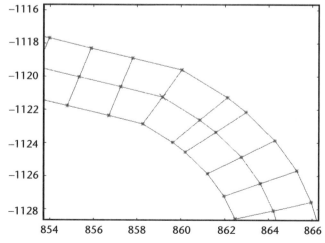

Figure 4.48   Road model with trapezoid corrections.

stretches regardless of how the original map database waypoints that define that lane were constructed. These adjacent lanes will be used in cases where passing or U-turns are required. A sample result is shown in Figure 4.49. This strategy is preferable to simply using the lane center waypoints for the adjacent lane, as they may not align with those of the existing lane and various anomalies, such as overlapping lane stretches or empty areas between lane stretches, may arise. Finally, one must consider situations when passing is not allowed.

After these road stretches are created we can begin looking for relevant obstacles in our lane. Generally, it is convenient to construct only stretches close to the vehicle and to repeat the process as the vehicle travels down the road.

Generating a road as a set of polygons allows extension of the current implementation with information coming from different sources. This compares favorably with many algorithms proposed in the literature to extract the edges of lanes [4, 28, 29] with a prescribed accuracy.

4.6.3   Intersection Generation

Intersection shape generation is related to lane generation. When the algorithm sees a waypoint marked as an intersection in its current path it requests the lane waypoints to each side of the intersection from the path planner, which has access to the entire map database. The path planner can identify all of the entry and exit points in the vicinity of the possible routes and can determine the center of the intersection by taking an average of these points. The intersection center point and the entry point of the route can be used to determine the location of the other lanes with respect to the direction of travel: right, front, left, and adjacent. For three-way intersections, one of these directions will be empty. The waypoints of the lanes can be used to generate stretches for all of the lanes that form the intersection and then an intersection shape can be created by extending the rightmost stretch of a road and the leftmost stretch of a road that is positioned counterclockwise to it. The intersection shape shown in Figure 4.50 was generated in this fashion. For a three-way intersection,

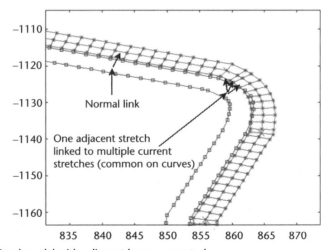

Figure 4.49 Road model with adjacent lanes generated.

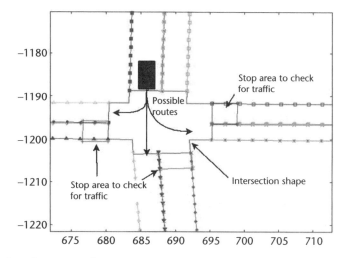

Figure 4.50   Sample generated intersection model.

an intersection may not be found if the extensions are parallel. If an intersection cannot be found, then the average of the stretch corner points can be used.

4.6.4   Primitives

Some computational geometric primitives are applicable to the situation analysis task. To deal with many situations, we must be able determine if an obstacle is inside a lane. To handle this case, algorithms for finding points inside polygons have been carefully taken into account. Many of these algorithms can be found in the computer graphics literature [30–32]. Most of them are very simple to implement, and have a low computational cost, typically $O(n\log n)$ or $O(n^2)$, with n the number of the vertices in the polygon.

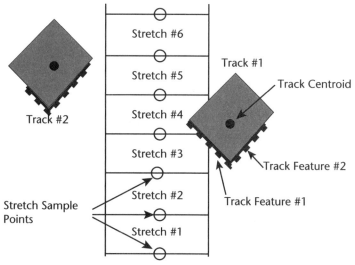

Figure 4.51   Stretches and target tracks to be evaluated.

As we define the target cluster geometry with a set of lines, one may also want to check if at least one line is inside a polygon. One approach to this problem is to determine whether any of the line segments composing the track intersect (cross) the boundary of the polygon. We must identify the case where an entire line segment is within the polygon by checking for endpoints lying inside the polygon as described above. As part of this operation, we can determine which side of the polygon the track line crosses or is near and the normal distance from the edge of the polygon.

(a)

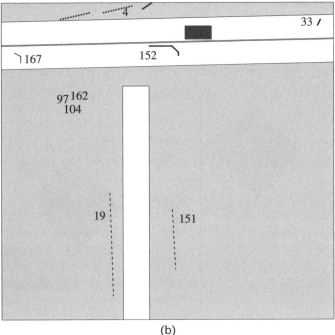

(b)

Figure 4.52   (a–d) First left turn in area A tests.

4.6.5 Track Classification

After road stretch polygons for the current path have been constructed, they can be used to determine if an obstacle is on this path. Each track contains a unique ID, a center location, a velocity, and track features. Track features are the two end points of a line segment. In Figure 4.51 two target tracks are shown: one off the road and another on the road. The rectangle is the actual object, the dots are from the LIDAR scan, and the lines are the track features. To determine if these objects are

(c)

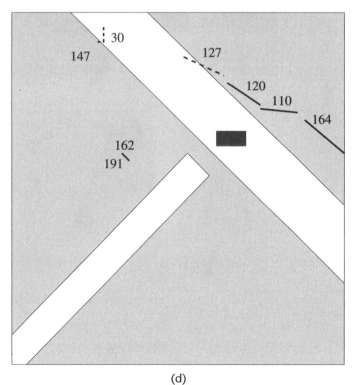

(d)

Figure 4.52 (continued)

on the road, one can first iterate through each track and determine if its centroid is within the stretch's polygon or the centroid is within 4 meters of the stretch's sample points (the two points at the center of the road). Four meters were chosen in the deployed ACT algorithm since it is a typical car length and double a stretch length; this means that a stretch will not associate with a track that has a centroid more than two stretches away. After linking a relevant track to a stretch number, one can determine if the track is a blocking or nonblocking obstacle.

To do this the situation analysis algorithm considers the track features and lines. The simplest case is when the track contains no lines, only a centroid. In this case, the normal to the right side of the stretch that passes through the centroid can be calculated and the length of this normal considered as the offset distance. In addition, the distance from the autonomous vehicle's current position to the obstacle's position can be computed by iterating from the current stretch to the ob-

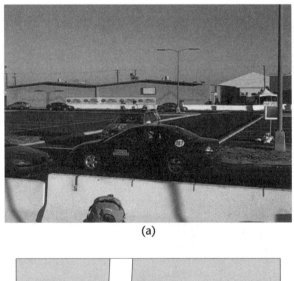

(a)

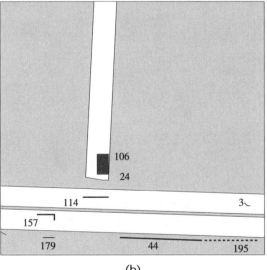

(b)

Figure 4.53 (a–f) Second left turn in area A tests.

stacles stretch, adding their lengths together, and finally adding the distance from the obstacle's stretch to the obstacle's centroid.

If the track has line segments, one can determine if the lines intersect the left or right edges of the stretch. If the track's line segments intersect both sides of a stretch, this track is identified as a blocking obstacle with an offset distance equal to the lane width and the distance to the current vehicle position is calculated as described above. If the track's line segments intersect one of the stretches' sides, one can determine which side it crosses and if that line crosses the front or back of the stretch. If the track's lines intersect with one side of the stretch but do not intersect with the front or back, then one can determine which endpoint is contained within the stretch and calculates the normal distance from the side that is intersected to the endpoint (similar to the no line case). If the track's lines cross a side and the front

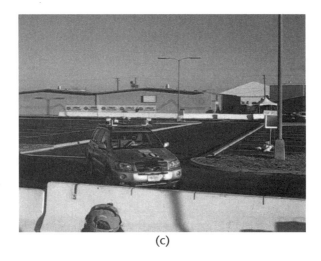

(c)

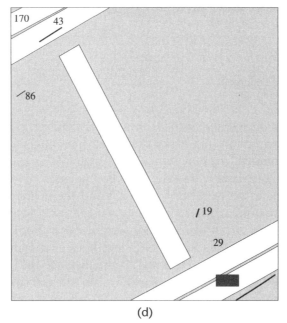

(d)

Figure 4.53   (continued)

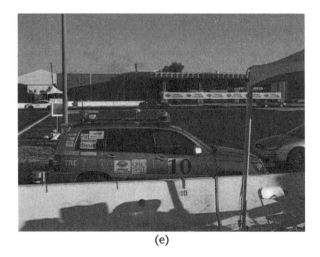

(e)

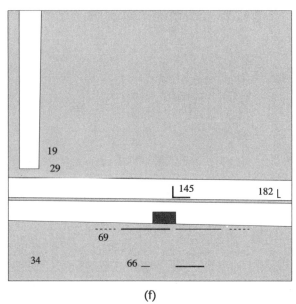

(f)

Figure 4.53   (continued)

or back, the intersection to the front or back can be treated like the endpoint in the previous case. Finally, if none of the track's lines intersect, one can determine if the lines are contained entirely within the stretch by testing to see if both endpoints of the line are contained within the stretches' polygon.

After the offset distance of a track and the distance to the object are calculated, the situation analysis algorithm will continue classifying the track in the next stretch. A larger offset distance dominates. For example, in Figure 4.51, track #1 would first be classified in stretch 3 and later in stretch 4. However, the distance to the obstacle would be recorded as the distance to the intersection in stretch 3 so that the high-level control algorithms can attempt to dodge the obstacle, if possible, and reach a safe offset distance well before the maximum offset distance of the obstacle is reached.

4.6.6   Sample Results

To illustrate the situations encountered by the autonomous vehicle and the resulting data with an example, we select a few instances from a run of the DARPA Urban Challenge designated area A. In this test, a large number of manually driven vehicles are circling a two-lane test track in both directions. The objective for the autonomous vehicle is to complete as many circuits of the track as possible in 30 minutes. In order to accomplish this, the vehicle must make two left-hand turns. The first, shown in Figure 4.52, requires crossing a single lane of oncoming traffic. The second, shown in Figure 4.53, requires stopping and waiting for an opportunity to cross one lane of traffic and merge into the far lane of traffic. Penalties were deducted if either the autonomous vehicle failed to move when an opening was available or moved in spite of the presence of opposing traffic. Although the traffic vehicles were manually driven and their drivers attempted to avoid collisions, a number of other teams did experience impact collisions or emergency stop conditions during this test.

Figure 4.52 shows the first left turn, crossing a single lane of traffic. The top images show the situation visually, and the lower plots show the sensory data clustered and tracked, which includes both other vehicles and barriers and other objects on the sides of the roads.

Figure 4.53 shows images of the second left turn along with classified sensor data. In addition to the other vehicles and roadside objects, several occluding elements are present (the speed indicator and a light pole), which can be clearly seen in the sensor output.

References

[1]   http://www.navcen.uscg.gov.

[2]   http://www.faa.gov/about/office_org/headquarters_offices/ato/service_units/techops/navservices/gnss/gps/.

[3]   Redmill, K. A., "A Simple Vision System for Lane Keeping," *Proc. 1997 IEEE Conference on Intelligent Transportation Systems*, Boston, MA, November 1997, pp. 212–217.

[4]   Redmill, K. A., et al., "A Lane Tracking System for Intelligent Vehicle Applications," *Proc. 2001 IEEE ITSC*, Oakland, CA, August 25–29, 2001, pp. 273–279.

[5]   Bertozzi, M., et al., "Artificial Vision in Road Vehicles," *Proceedings of the IEEE*, Vol. 90, No. 7, July 2002, pp. 1258–1271.

[6]   Farkas, D., et al., "Forward Looking Radar Navigation System for 1997 AHS Demonstration," *Proc. 1997 IEEE Conference on Intelligent Transportation Systems*, Boston, MA, Nov. 1997, pp. 672–675.

[7]   Lewis, F. L., L. Xie, and D. Popa, *Optimal and Robust Estimation: With an Introduction to Stochastic Control Theory*, Boca Raton, FL: CRC Press, 2008.

[8]   Thrun, S., W. Burgard, and D. Fox, *Probabilistic Robots*, Cambridge, MA: MIT Press, 2006.

[9]   Redmill, K. A., T. Kitajima, and Ü. Özguner, "DGPS/INS Integrated Positioning for Control of Automated Vehicles," *Proc. 2001 IEEE Intelligent Transportation Systems Conference*, August 25–29, 2001, pp. 172–178.

[10]   Xiang, Z., and Ü. Özguner, "A 3D Positioning System for Off-Road Autonomous Vehicles," *Proc. 2005 IEEE Intelligent Vehicle Symposium*, June 6–8, 2005, pp. 130–135.

[11] Martin, M. C., and H. P. Moravec, *Robot Evidence Grids,* Carnegie Mellon Robots Institute Technical Report CMU-RI-TR-96-06, 1996.

[12] Hall, D. L., and J. Llinas, "An Introduction to Multisensor Data Fusion," *Proceedings of the IEEE,* Vol. 85, No. 1, January 1997, pp. 6–23.

[13] Bar-Shalom, Y., and X. Li, *Multitarget-Multisensor Tracking: Principles and Techniques,* University of Connecticut, Storrs, CT, 1995.

[14] Foresti, G. L., and C. S. Regazzoni, "Multisensor Data Fusion for Autonomous Vehicle Navigation in Risky Environments," *IEEE Trans. on Vehicular Technology,* Vol. 51, No. 5, 2002, pp. 1165–1185.

[15] Redmill, K. A., J. Martin, and Ü. Özguner, "Sensing and Sensor Fusion for the 2005 Desert Buckeyes DARPA Grand Challenge Offroad Autonomous Vehicle," *Proc. IEEE Intelligent Vehicles Symposium,* June 13–15, 2006, pp. 528–533.

[16] Hummel, B., et al., "Vision-Based Path-Planning in Unstructured Environments," *Proc. IEEE Intelligent Vehicle Symposium,* June 2006, pp. 176–181.

[17] Dang, T., and C. Hoffman, "Fast Obstacle Hypothesis Generation Using 3D Position and 3D Motion," *Proc. IEEE International Workshop on MachineVision for Intelligent Vehicles,* June 2005.

[18] Abrash, M., *Graphical Programming Black Book,* Coriolis Group Books, 1997, http://archive.gamedev.net/reference/articles/article1698.asp.

[19] Galler, B. A., and M. J. Fischer, "An Improved Equivalence Algorithm," *Communications of the ACM,* Vol. 7, No. 5, May 1964, pp. 301–303.

[20] MacLachlan, R., *Tracking Moving Objects from a Moving Vehicle Using a Laser Scanner,* Carnegie Mellon Technical Report CMU-RI_TR-0507, 2005.

[21] Jonker, R. and A. Volgenant, "A Shortest Augmenting Path Algorithm for Dense and Sparse Linear Assignment Problems," *Computing,* Vol. 38 , 1997, pp. 325-340.

[22] Benedict, T. R., and G. W. Bordner, "Synthesis of an Optimal Set of Radar Track-While-Scan Smoothing Equations," *IRE Transactions on Automatic Control,* Vol. AC-1, July 1962.

[23] Nagel, H. H., "Steps Toward a Cognitive Vision System," *AI Magazine,* Vol. 25, No. 2, 2004, pp. 31–50.

[24] Albus, J., et al., "Achieving Intelligent Performance in Autonomous Driving," National Institute of Standards and Technology, Gaithersburg, MD, October 2003.

[25] Catmull, E. E., and R. J. Rom, "A Class of Local Interpolating Splines," in *Computer Aided Geometric Design,* R. E. Barnhill and R. F. Riesenfeld, (eds.), Orlando, FL: Academic Press, 1974, pp. 317–326.

[26] Barry, P., and R. Goldsman, "A Recursive Evaluation Algorithm for a Class of Catmull-Rom Splines," *Computer Graphics, (SIGGRAPH '88),* Vol. 22, No. 4, 1988, pp. 199–204.

[27] Wang, Y., D. Shen, and E. K. Teoh, "Lane Detection Using Catmull-Rom Spline," *Proc. 1998 IEEE International Conference on Intelligent Vehicles,* 1998.

[28] Schreiber, D., B. Alefs, and M. Clabian, "Single Camera Lane Detection and Tracking," *Proc. 2005 IEEE Intelligent Transportation Systems,* 2005.

[29] Kim, S., S. Park, and K. H. Choi, "Extracting Road Boundary for Autonomous Vehicles Via Edge Analysis," *Signal and Image Processing SIP 2006,* Honolulu, HI, July 2006.

[30] Haines, E., "Point in Polygon Strategies," *Graphics Gems IV,* P. Heckbert, (ed.), New York: Academic Press, 1994, pp. 24–46.

[31] Shamos, M. I., and F. P. Preparata, "The Standard but Technically Complex Work on Geometric Algorithms," in *Computational Geometry,* New York: Springer-Verlag, 1985.

[32] O'Rourke, J., "Section 7.4, Point in Polygon," in *Computational Geometry in C,* 2nd ed., New York: Cambridge University Press, 1998.

CHAPTER 5

Examples of Autonomy

5.1 Cruise Control

Actual and modern cruise control consists of advanced technology applications to in city driving and highway systems in order to enhance mobility and safety, and to increase traffic flow by reducing possible congestion through using advanced electronics, communications, and control technologies. Former cruise control systems or longitudinal control of automated vehicle systems have worked well on open highways or open freeways while regulating the speed of the vehicle at the driver's set point. When the traffic was congested or a slow car ahead was perceived by the driver, the cruise systems were disengaged and the speed was changed manually by the driver's intervention to the throttle pedal or the brake pedal. Recently adaptive cruise control (ACC) systems have been introduced as a technological improvement over existing cruise controllers on ground vehicles. ACC systems regulate the vehicle speed to follow the driver's set point if there is no vehicle or any obstacles in sight. When a slower vehicle is observed ahead, the ACC controlled vehicle will follow the vehicle ahead at a safe distance by adjusting its relative speed. Now, ACC systems are capable of maintaining a controlled vehicle's position relative to the leading vehicle including in congested traffic and even in city traffic by using stop-and-go features while maintaining a safe distance between leading and following vehicles autonomously. The current cruise technology helps the driver by taking his or her driving load.

Intelligent cruise may reduce the probability of accidents while sharing the workload of the human driver and giving him or her some mental comfort to consider other situations during the trip. Automation improves traffic flow by using lanes more efficiently and by setting appropriately safe headway distances among the leading and following cars. As an additional benefit, the automated systems eliminate unnecessary acceleration and deceleration in the open roads to minimize fuel consumption and, emission and maximize engine efficiency.

5.1.1 Background

Cruise control is basically a throttle feedback loop, whereas ABS is a brake control feedback loop. From the pure control viewpoint, the realization of an intelligent vehicle would require two additional capabilities: the ability to jointly control multiple loops and the ability to close the third loop, steering.

In general, studies on the design of automated vehicle system involve the solution of two decoupled control systems: the steering control (lane keeping) and the longitudinal (headway) control. The longitudinal control (i.e., speed control—cruising at the selected speed or keeping relative speed and relative position of the controlled vehicle with respect to the lead vehicles at a safe distance in the highway traffic) constitutes the basis for current and future advanced automated automotive technologies. Development of a longitudinal controller for headway regulation and a supervisory hybrid controller capable of switching among different control actions was proposed by Hatipoglu, Özgüner, and Sommerville [1]. The longitudinal car model was built by assembling submodels of the engine, torque converter, transmission, brake, and vehicle body dynamics in the longitudinal direction. The headway controller was presented by introducing the dynamics of the relative velocity, the relative distance, and also the highway safety distance. A controller with constant acceleration (or deceleration), smooth acceleration (or deceleration), and linear state feedback was presented towards an intelligent cruise control (ICC) hybrid model. ICC hybrid modeling with the decision rules and boundaries, and phase plane analysis of relative distance versus relative velocity in different control regions, were introduced. The Kalman filter and its update law on a real-time basis was designed to estimate the relative distance and velocity sensed by a laser rangefinder. Possible discrete jumps due to the quantized sensor reading and the process noise were eliminated by the designed filter. Control methodology including collision region and conventional cruise control when no obstacle/vehicle is detected in the sensing range was validated by simulation and experimental results [1].

Intelligent cruise control theory and its applications have been investigated by many researchers. Using the dynamics of the relative velocity and relative distance, PID controller with fixed gains, gain scheduling of the PID controller, and an adaptive controller were considered for headway regulation by Ioannou and Xu [2]. Basically, transfer function between the deviation of the vehicle speed and the throttle angle was approximated for the operating point at steady-state, and the dynamics of the relative speed and distance of the leading and following vehicles were driven to zero by using the considered controllers. The feedback control for the braking dynamics was proposed by using feedback linearization and the effects of brake torque, static friction forces, and rolling friction forces and aerodynamic forces were suppressed. Switching logic between the brake and throttle controller was presented by dividing operating region of the vehicles into three possible situations. Vehicle following in a single-lane without passing the leading vehicle was studied by Ioannou and Chien [3]. Replacement of the human driver by an autonomous cruise controller was shown to enhance traffic flows. The different driver models were considered for stand-alone cruise controller applications where no information was exchanged between vehicles but all of the vehicles were equipped by the proposed cruise controllers. Cooperative adaptive cruise controller design when intervehicle communication was added to adaptive cruise controller was investigated by Lu and Hedrick [4]. Relative distance and velocity with respect to the lead vehicle measured by radar were used as the variables to be regulated by the considered dynamic sliding mode controllers.

The challenges of cruise control systems arise from the complicated and complex mechanical, electromechanical, or electrical actuator systems. Vehicle actuator systems are throttle and engine controller systems, brake systems, and transmission control systems. These systems are in general nonlinear involving nonsmooth nonlinearities in the input-output or in the derived analytical models. The implementation and comparison of nonlinear controllers in the longitudinal regulation of cars for platooning was presented by Lu, Tan, and Hedrick [5]. Real-time tests were accomplished by restructuring original Demo '97 codes developed by the PATH program with considered nonlinear control methodologies. A comparison of spacing and headway control methodologies for automated vehicles in a platoon were examined by Swaroop et al. [6]. A three-state lumped parameter longitudinal model of a vehicle was presented before detailed analysis of constant spacing control and constant headway control. Individual vehicle stability was shown to be established straightforward where string stability in a platoon was shown to impose intervehicle communication for constant spacing control. Constant headway control was capable of offering string stability without intervehicle stating of the relative position of each vehicle with respect to the lead vehicle. Shladover reviewed the state of development of advanced vehicle control systems in [7]. The contributions of The Ohio State University on both steering and longitudinal (spacing) control under the sponsorships of the Ohio Department of Highways and the Federal Highway Administration, PATH (Partner for Advanced Transit and Highways) program University of California, Berkeley and Personal Vehicle System (PVS), and Super-Smart Vehicle Systems (SSVS) under the sponsorship of Ministry of International Trade and Industry, Japan were stated. Tsugawa et al. presented an architecture for cooperative driving of automated vehicles by using intervehicle communications and intervehicle gap measurement. Proposed architecture and its layers were investigated in detail followed by experiments on a real-time basis [8].

5.1.2 Speed Control with an Engine Model

Cruise control or longitudinal speed control of automated vehicles has been simply a closed-loop setup to follow the desired constant speed reference generated by the human driver and enforced by the speed controller acting on the throttle angle or pressure of the brake actuator. Closed-loop vehicle longitudinal controller for the reference speed tracking is illustrated in Figure 5.1 constituted by the dynamics of the models. In Figure 5.1(a), the vehicle model as a point mass is presented for closed-loop speed controller development. The point mass model is given by $m\ddot{x} + \alpha\dot{x} = f$ [see also (2.1)] and a feed forward longitudinal controller (probably a simple PI controller) is added in the closed-loop. Figure 5.1(b) shows a more complex longitudinal vehicle model involving nonlinear models of the engine, torque converter, and transmission; the internal couplings among those models is plotted to illustrate the complexity of the closed-loop longitudinal speed control scheme. In this figure α is the throttle angle, w_e is the engine rotational speed, T_t, T_p, and T_s are the turbine, pump, and shaft torques, respectively, and w_t is the turbine rotational speed [1]. In this chapter, notation is slighlty changed to introduce more states and

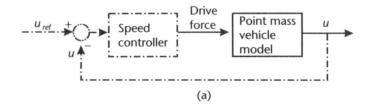

(a)

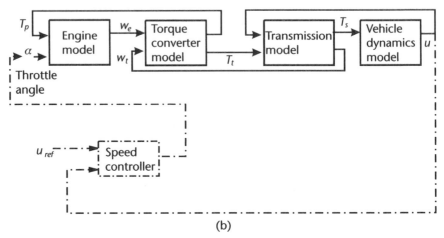

(b)

Figure 5.1 (a) Point mass simplified vehicle model, and (b) longitudinal vehicle model for cruise control [1].

variables. The velocity in the longitudinal and lateral direction, which is given by $\dot{x}$ and $\dot{y}$ in Chapter 2, is replaced by its equivalent symbol u and v, respectively (i.e., $\dot{x} \equiv u$, $\dot{y} \equiv v$).

A linear longitudinal speed controller is derived by using the vehicle acceleration according to Newton's law,

$$M\dot{u} = F - F_{load} \tag{5.1}$$

where F denotes the net force generated by the engine model and transformed by the transmission model following the switching strategies as function of the actual and requested speed values. F_{load} represents aerodynamic drag forces and possible load force created by the road inclination. It may be written as a sum of two load forces resisting to the force created by the engine:

$$F_{load} = \gamma u^2 + gM \sin(\alpha) \tag{5.2}$$

The force accelerating the vehicle may be converted into the generated engine torque in terms of the transmission gain as $F = \dfrac{K_i}{R_e} T_e$ where K_i, $i = 1, 2, 3$, is the gain of the transmission, R_e denotes the effective radius of the wheel, and T_e denotes the torque generated by the engine model. The vehicle speed is derived in terms of engine torque, transmission gain, road, and aerodynamic loads,

$$\dot{u} = \frac{K_i}{R_e M} T_e - \frac{\gamma}{M} u^2 - g\sin(\alpha) \tag{5.3}$$

The simple speed controller may be achieved by using a proportional-integral (PI) controller, and generating the engine torque in order to create and maintain zero error value between the reference speed value given by the driver and the actual vehicle speed.

The engine torque term as the output of the PI controller, denoted by T_e, is given as

$$T_e = K_p \left(u_{ref} - u\right) + \frac{K_p}{T_I} \int_0^t \left(u_{ref} - u\right) d\tau + \frac{R_e}{K_i} \gamma u^2 \tag{5.4}$$

The closed loop speed dynamics becomes

$$\dot{u} = \frac{K_i}{R_e M} \left(K_p \left(u_{ref} - u\right) + \frac{K_p}{T_I} \int_0^t \left(u_{ref} - u\right) d\tau \right) - g\sin(\alpha) \tag{5.5}$$

achieving zero steady-state error, $e \equiv u_{ref} - u$, between the reference and actual speed of the vehicle. The choice of controller parameters K_P and T_I are tuned to obtain the desired transient performance. The steady-state error is guaranteed to be zero through the integral term in the PI controller. The transient performance exhibited during the acceleration or deceleration maneuver of the vehicle model may be considered important to satisfy ride comfort, driving performance, and also fuel consumption economy. The integral time constant T_I may be tuned for a fast transient response without exceeding the reference value. If a large value is chosen, slow transient may occur, and if a small value is chosen, a fast transient with an overshoot followed by an oscillatory response may be generated in the time responses of the vehicle speed. On the other hand, the choice of proportional gain, K_p, is chosen in order to assure desired change in the time response of the vehicle speed and not exceeding the service limits of 0.2g acceleration with a jerk 0.2 g.sec^{-1} for ride comfort and fuel consumption economy. Higher gain may cause higher acceleration or deceleration.

The generated engine torque is to accelerate the vehicle model. In order to decelerate the vehicle dynamics (i.e., when the driver's set point for vehicle speed is reduced), the vehicle speed is regulated to the new lower reference speed value and braking torque is replaced by engine torque. To achieve smooth braking transient performance, once again different controller gains may be chosen since between the brake and vehicle model, transmission or torque converter model do not exist. The performance criteria remains the same, the service limits of −0.2g deceleration with a jerk −0.2 g.sec^{-1} for ride comfort may be achieved while the driver does not intervene to the longitudinal controller of the vehicle model.

5.1.2.1 Engine Model

The engine model is by far the most complex part of the longitudinal part of the longitudinal car dynamics. An accurate engine model requires many states and must include descriptions of the intake and exhaust air flow dynamics, the discrete combustion occurrences in the cylinders, the crank shaft dynamics, the air-fuel ratio, the spark advance, and many other nonlinear effects. One of the first simplifications has been to threat the engine as a mean flow, which only models the average air flow and torque production. The most commonly used mean flow model is described by Cho and Hedrick [9]. An alternative approach has been to simplify the model even more by incorporating most of the engine dynamics in a static three-dimensional map which relates throttle angle and engine angular speed to steady-state output torque. By using this map, the engine model can be simplified to a single state representing engine speed. Engine combustion torque can be derived in terms of engine angular speed and throttle angle, $T_c = f(w_e, \Theta)$ as illustrated in Figure 5.2.

The differential equation describing the engine model in terms of throttle angle input and engine angular speed

$$\dot{w}_e = \frac{1}{J_e}\left(T_c\left(\Theta + \Theta_{offset}, w_e\right) - T_a\left(w_e\right) - T_p\left(w_e, w_t\right)\right) \tag{5.6}$$

where

Θ: throttle angle;

w_e: engine speed;

w_t: torque converter turbine speed;

J_e: engine inertia;

T_c: engine combustion torque;

T_a: engine accessory and load torque;

T_p: feedback from torque converter pump.

The nonlinear engine model may be given also by a static linear equation in terms of throttle angle input and engine angular speed affecting the output of the engine model by different constants depending on the operating regions where the torque equality is linearized (see Figure 5.2 for a static linear engine torque-speed equation as a parameter of throttle angle input).

An analytical equation for the linearized characteristics of engine combustion with respect to engine angular speed and throttle angle is given as

$$T_c(w_e, \Theta) = C_1\Theta - C_2 w_e \tag{5.7}$$

where C_1 and C_2 denote the local slopes of engine torque-angular speed region with respect to different throttle angle variations as illustrated in Figure 5.3 on the two-dimensional plane.

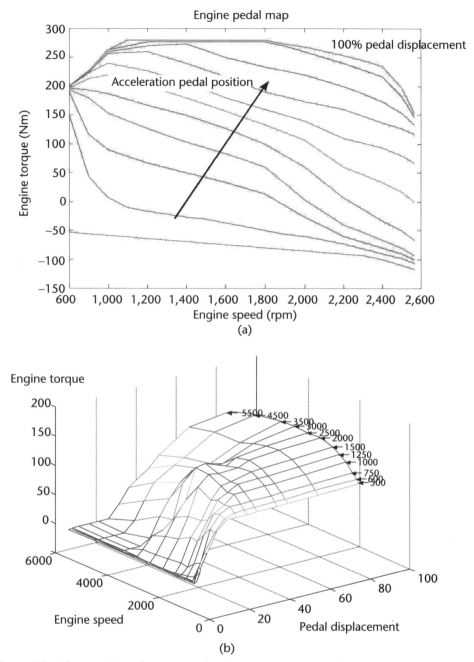

Figure 5.2 Characteristics of engine combustion torque in terms of engine angular speed versus throttle angle. (a) Torque-speed characteristics versus the pedal displacement. (b) Torque, speed, and pedal displacement characteristics.

5.1.2.2 Torque Converter Model

A torque converter is used in vehicles with automatic transmission to provide the coupling between the engine and the transmission. Three useful properties of a torque converter are: providing torque amplification during the initial acceleration from a stop, smoothing the torque pulsations created from the individual cylin-

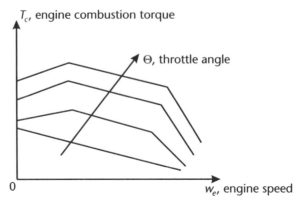

Figure 5.3   Linearized characteristics of engine combustion torque in terms of engine angular speed versus throttle angle.

der firings in the engine, and smoothing the power transfer despite a varying load torque due to road variations such as bumps and potholes.

Two quadratic equations relating pump and turbine speed to pump and turbine torque are used to model the torque converter [10]. These equations are linearized at the operating pump and turbine speeds, w_{c0} and w_{T0}, respectively.

$$T_p = 2m_1 w_{e0} we + m_2 w_{e0} w_T + 2m_3 w_{T0} w_T$$
$$T_T = 2n_1 w_{e0} we + n_2 w_{e0} w_T + 2n_3 w_{T0} w_T$$

where w_T denotes turbine speed (rpm), T_T denotes turbine torque (ft lbs), T_P is pump torque (ft lbs), and m_i, n_i are constant coefficients. Linearized torque converter model at the turbine and engine operating speeds can be used for the proposed linear longitudinal car model. The torque converter operates in three defined modes depending on the ratio of turbine and engine speed. These modes are converter, coupling, and overrun. The operation is determined by the speed ratio

$$SR = \frac{w_T}{w_e}$$

The torque converter is considered to be in converting mode if $SR < 0.842$, coupling mode if $0.842 \leq SR \leq 1.0$, and overrun mode if $SR > 1.0$. These upper and lower limits of the speed ratios (SR) define the hybrid operation of the torque converter.

5.1.2.3   Transmission Model

Three different gear ratios are considered for the hybrid longitudinal car model. Assuming the gears are lossless and frictionless, a constant gear ratio is used to describe the torque transfer from the torque converter to drivetrain. Transmission of gear ratios are considered to be three different gains K_1, K_2, and K_3, holding the inequalities $K_1 < K_2 < K_3$ where K_1 corresponds to the gain of the first gear and so on.

The hybrid transmission model considers the different gains from the discrete set $\{K_1, K_2, K_3\}$ and uses the continuous vehicle speed and throttle angle. When the vehicle speed u exceeds the upper and lower limits represented by Figure 5.4 by the notations u_{0i}^+, u_{0i}^-, respectively, the change of gear or mathematically discrete gain change occurs denoted by K_i, $i = 1, 2, 3$. The switching logic is achieved by defining the switch sets;

$$K_{i,i+1} = \left\{ u \geq u_{0i}^+, \theta_{min} \leq \theta \leq \theta_{max} \middle| u = \frac{R_e}{K_{i+1}} w_e \right\} \text{ for } i = 1, 2$$

$$K_{i+1,i} = \left\{ u \leq u_{0i}^-, \theta_{min} \leq \theta \leq \theta_{max} \middle| u = \frac{R_e}{K_i} w_e \right\} \text{ for } i = 1, 2$$

where R_e is the effective radius of tire and defines transformation between angular velocity and longitudinal velocity of the vehicle $R_e w_e = u$.

The longitudinal model involving all different modes for torque converter and different gear gain for the transmission models are illustrated in Figure 5.5. The engine model and vehicle dynamics are assumed to be operating at linearized pump torque, T_{P0}, turbine torque, T_{T0}, at the operating angular speeds for operating pump and vehicle speed, w_{c0} and w_{v0}, respectively.

Speed control with an engine model may be represented by continuous vehicle dynamics model where the vehicle velocity and engine angular velocity variables are continuous with respect to time described by differential equations (5.1) and (5.6), and by the discrete event systems where dynamics of the torque converter model and transmission systems are changed discretely in time depending on the upper and lower limits of the vehicle and engine speed. Speed control with an engine model is illustrated in Figure 5.5, where each model interacts with other continuous or discrete models and the overall system response is affected by the interaction and coupling of these different type of systems dependent on the hybrid limits.

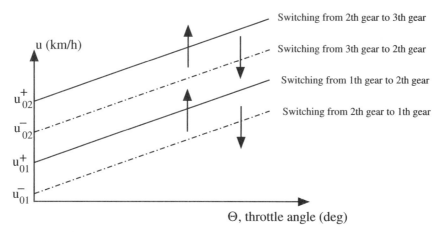

Figure 5.4 Simplified characteristics of automatic transmission model and switching logic presenting the hybrid operation in terms of vehicle speed versus throttle angle.

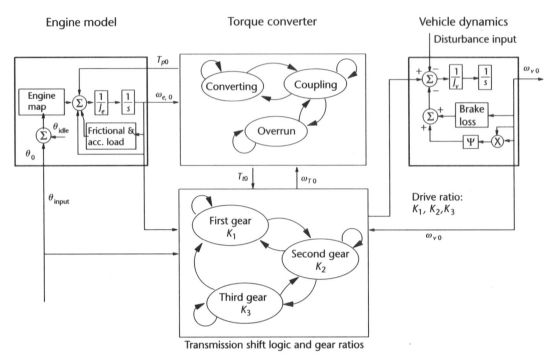

Figure 5.5 Overview of vehicle dynamics, engine model coupled with torque converter, and transmission illustrated with different operating modes.

5.1.3   More Complex Systems

Although both ABS and cruise control have attributes of autonomy, one point of transition from *standard* vehicles to so-called *intelligent* vehicles relates to the ability to slow down in cruise control mode, when a vehicle detects a slower vehicle ahead. This feature is referred to as advanced cruise control (ACC) and cruise assist systems or, less popularly, as adaptive cruise control and intelligent cruise control.

ACC implies that there exists a means by which a vehicle ahead can be detected, its distance measured, and its relative velocity measured or calculated. Various technologies exist to do this at this time. Popular among them are laser and radar systems, although vision-based techniques can also be considered. ACC can be used either for cars or trucks, or it can be part of a full convoy operation (where lateral control is also implied).

In this section, we are concerned with the longitudinal control aspects of the convoying problem. For the time being, it is assumed that the convoy consists of only two vehicles (trucks) in a leader-follower configuration as depicted in Figure 5.6, although the generalization to more vehicles is straightforward. The follower has to be able to detect the leader, estimate the distance to it, and estimate the leader's speed. Vision-, radar-, or laser-based systems can be utilized to accomplish these tasks. Fiducial tracking senses the mobile moving target objects, (see for instance [11, 12]). In Figure 5.7 we have indicated a *patch*, providing the follower's (forward-looking) sensor system an easily detectable target [13].

The main task of a longitudinal controller is to keep the distance between the vehicles at a desired safety level. Let the longitudinal velocity of the follower and

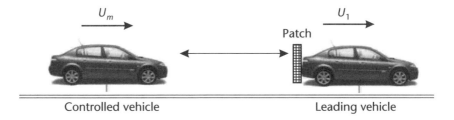

U_m : speed of the follower
U_1 : speed of the leader

Figure 5.6 Leader-follower vehicle setup.

Figure 5.7 The OSU visual patch.

the leader be U_M and U_L, respectively. The measured headway d and the safety distance d_s can be defined as

$$d = x_L - x_M$$
$$d_s = hU_L + d_0 \tag{5.8}$$

where X_L and X_M are the longitudinal positions of the leader and the follower, respectively, h stands for headway (the time it takes for the leader to stop), and d_0 provides an additional safety margin. The velocity difference ΔU is given by:

$$\Delta U = U_M - U_L = -\dot{d} \tag{5.9}$$

Consider Figure 5.8. The strategy for regulation is as follows: The recommended velocity of the follower should be chosen in such a way that the velocity vector of the solution trajectory in the $(d, \Delta U)$ plane is directed to the $(d_s, 0)$ point at any time. This choice enforces state trajectories toward the goal point on a straight line whose slope is determined by the initial position of the system in the $(d, \Delta U)$ plane and guarantees that the velocities of the vehicles become equal when the desired safety distance is achieved. The slope of the line on which the trajectory slides to the goal point determines the convergence rate. We divide the $(d, \Delta U)$ plane into six regions. The collision region and the constant velocity region are also included in Figure 5.8. In the constant velocity region, the follower keeps its current velocity until the distance between the vehicles becomes less than a user-defined critical distance d_c. Figure 5.8 also includes a relative acceleration curve that gives the least possible value of d at which the follower should begin to decelerate at its maximum rate to be able to reach the goal point for a given ΔU, assuming a constant velocity for the leader. The figure also includes a minimum convergence rate line (MCRL) whose slope is chosen by considering the minimum admissible convergence rate.

In region 2 and region 5, it is physically impossible to enforce the trajectories toward the goal point on a straight line. So, the controller should decelerate (accelerate) the follower at the maximum deceleration (acceleration) rate in region 2 (region 5). In region 3 and region 6, it is possible to steer the trajectories to the $(d_s, 0)$ point through a straight line between the initial point and the goal point. However, the convergence rate would be smaller than the minimum admissible convergence rate because the slope of the line is less than the slope of the MCRL. So, in region 6 (region 3) we prefer first accelerating (decelerating) the follower toward the MCRL at its maximum acceleration (deceleration) rate and then sliding the trajectories to the goal point through this line. In region 1 and region 4, the desired velocity can be calculated as follows:

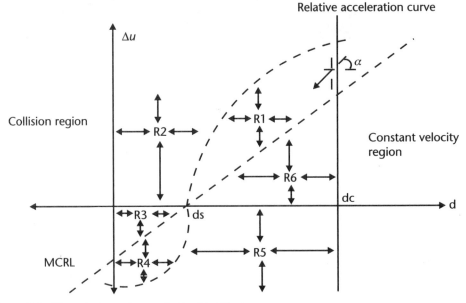

Figure 5.8 The solution trajectory in the $(d, \Delta U)$ plane.

$$m = \tan(\alpha)$$

$$m = \frac{\Delta \dot{U}}{\dot{d}} = -\frac{a_M - a_L}{\Delta U}$$

$$m_{des} = \frac{\Delta U}{d - d_s} \tag{5.10}$$

$$m = m_{des} \Rightarrow a_M = -\frac{(\Delta U)^2}{d - d_s} + a_L$$

where m is the slope of the trajectory velocity vector, m_{des} is the desired slope, and a_M, a_L are the accelerations of the follower and the leader, respectively.

Equation (5.10) gives the necessary acceleration for the follower that ensures the exact convergence of the solution trajectory to the goal point on a straight line. However, it may not always be possible to obtain this acceleration due to the acceleration and jerk limits of the vehicle. The bounds on the acceleration are determined by the physical capacity of the vehicle, whereas jerk limits are mainly determined by riding comfort. In the other regions, the above argument also holds except that a_M is taken as $a_{max}(a_{min})$ in region 6 and region 5 (region 3 and region 4) instead of using the equation.

5.2 Antilock-Brake Systems

5.2.1 Background

Antilock brake systems enhance vehicle safety, performance, and handling capabilities. Antilock brake systems (ABS) maintain directional stability and steerability under emergency maneuvering tasks on a slippery road surface. With developing digital electronics technology, for example, BOSCH has been able to offer the digitally controlled ABS to the automotive producers since 1978 and to the commercial vehicles market since 1981 [14].

During emergency braking on a slippery road surface, the wheels of the vehicle may lock; this phenomenon occurs since the friction force on the locked wheel is usually considerably less than the demanded braking force by the driver. The vehicle slides on the locked wheels increasing the stopping distance and more importantly may lose directional stability and steerability under most of the driving conditions and maneuvering tasks. The friction coefficient between tire patches and road pavement cannot be known by the average driver and this external parameter affecting the stopping distance and stopping time are changing with weather conditions, road pavement, and the type and quality of the tires. Due to the strong nonlinearity with time-varying parameters and uncertainties in the vehicle brake systems, and without a priori knowledge of the tire, road friction coefficient, or condition, design of ABS becomes a difficult control problem.

A discrete-time ABS controller was designed by using a linear control design with integral feedback subject to the locally linearized brake and vehicle system dynamics involving the lumped uncertainty parameter [15]. Linearization of highly nonlinear systems, lumped uncertain parameter approach, and assumption of the knowledge of vehicle velocity were the disadvantages of the design. Linear theory may not be a solution to the design of ABS. Some hybrid controller designs were addressed in [16]. The ABS hydraulic actuator has been considered to have discrete states {brake pressure hold, brake pressure increase, brake pressure reduction} and a finite state machine supervising ABS activation logic of feedforward and feedback controllers among the choices of several modes subject to the estimation of wheel angular speed [17]. Gain-scheduled ABS design was presented for the scenarios when the vehicle speed was a slowly time varying parameter and the brake dynamics were linearized around the nominal wheel slip [18]. To estimate tire-road friction on a real-time basis, an algorithm based on a Kalman filter was proposed in [19]. For automotive applications where uncertainties, discontinuities and lack of measured variables are major controller design problems, sliding mode control theory has been introduced to assure online optimization, estimation, friction compensation, and traction control in automotive control problems [20]. Sliding mode controller methodology was proposed to track an unknown optimal value even in the case that the value changes on a real-time basis. ABS control using the proposed optimal search algorithm assures reaching and operation at the maximum friction force during emergency braking without a priori knowledge of the optimal slip [21]. Time-delay effects caused by the hydraulic actuator dynamics were considered by using the extremum seeking approach in [22]. Oscillations occurring around the maximum point of the tire force versus slip characteristics were attenuated by using a higher-order sliding mode controller in [23].

5.2.2 Slip

The tire force model between the tire patches and the road surface depend on the road surface, tire, weather, and many other conditions that may not be known a priori (see Figure 5.9 for parameter-dependent characteristics). Friction force affecting the individual wheel rotational dynamics during braking is a nonlinear function of the relative slip [i.e., $F_i = F_i(k_i)$]. The slip is defined as the relative difference between the decelerated wheel's angular velocity multiplied by the effective radius of the tire, $R_e w_i$, and the velocity in the longitudinal direction, u;

$$k_i = \frac{R_e w_i - u}{u} \tag{5.11}$$

for the case of braking, when

$$R_e w_i \le u$$

and as

$$k_i = \frac{R_e w_i - u}{R_e w_i} \tag{5.12}$$

for the case of acceleration when

$$R_e w_i > u$$

for $i = 1, 2, 3, 4$. The values of the function $F_i(k_i)$ were obtained experimentally for different types of surface conditions. Experiments showed that in the region $k_i > 0$ the function has a single global maximum and in the region $k_i < 0$ a global minimum. The form of the function F_i for $k_i > 0$ is plotted in Figure 5.9. For different road types (i.e., dry, wet, or icy road conditions), a global maximum point of the tire force model is encircled.

A more general tire model, which does not contradict the Pacejka model but includes it as a particular case, has been considered in [24]. It is assumed that each friction force F_i is a nonstationary function of the slip k_i,

$$F_i = F_i(t, k_i)$$

with bounded partial derivatives

$$\left| \frac{\partial F_i}{\partial k_i} \right| + \left| \frac{\partial F_i}{\partial t} \right| \leq C_0$$

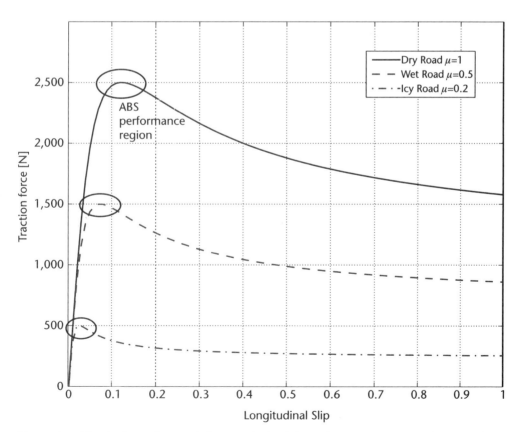

Figure 5.9 The tire force characteristics. Longitudinal slip with respect to traction force as a function of road condition.

and such that, for every t an inequality

$$k_i F_i(t, k_i) \geq 0$$

is maintained, and the function F_i has a unique global maximum at

$$k_i^*(t) \geq \delta > 0$$
$$y_i^* = F_i(t, k_i^*)$$

and a unique global minimum at

$$k_i^*(t) \leq -\delta < 0$$
$$y_{*i} = F_i(t, k_{*i})$$

For instance, a unique global maximum at the slip ratio k_i^* is plotted in Figure 5.10.

By assumption F_i is a sufficiently smooth function of k_i in the regions $k_i > 0$ and $k_i < 0$ and in the ε-vicinity ($\varepsilon > 0$) the extremal points k_i^* and k_{*i} satisfy the conditions;

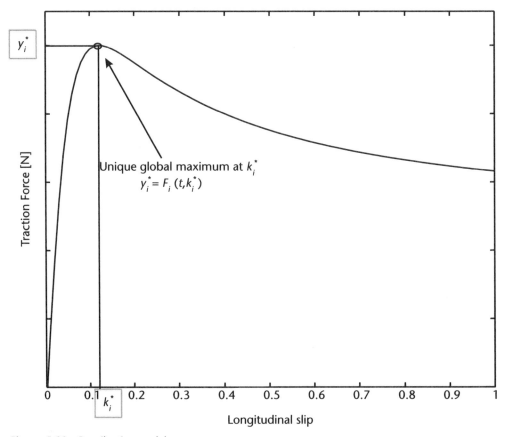

Figure 5.10   Pacejka tire model.

$$\left| \frac{\partial F_i(t, k_i)}{\partial k_i} \right| \geq K_0 \left| k_i - k_{*i} \right|$$

and

$$\left| \frac{\partial F_i(t, k_i)}{\partial k_i} \right| \geq K_0 \left| k_i^* - k_i \right|$$

The case when F_i depends on the absolute velocity v has been included allowing F_i to be nonstationary. It may be a function of the absolute slip

$$s = R_e w_i - u$$
$$F(t, k) = \tilde{F}(s) = \tilde{F}(v(t), k)$$

5.2.3   An ABS System

Antilock brake systems uses the brake hydraulic system dynamics model as an actuator. The common part of the system consist of a master cylinder, pump, and low reservoir. For each wheel there are two valves: a build valve, a damp valve, and a wheel cylinder. The valves are on/off type devices, and can be only in two positions: closed or open. The pressure created by the driver and the pump is transferred to the wheel cylinder only if the build valve is open and the dump valve is closed. If the build valve is closed and the dump valve is open, the pressure in the wheel cylinder decreases due to the fluid flow in the direction of the low pressure reservoir. The case when the build valve and the dump valve are open, is not allowed, but it is possible to have these valves closed at the same time. In this case, neglecting the fast transition process in the hydraulic line, the pressure in the wheel cylinder is assumed to be constant.

The flow in and out of the hydraulic circuit for the ith wheel cylinder can be modeled as a flow through an orifice:

$$Q_i = A_1 c_{d1i} \sqrt{\frac{2}{\rho}(P_p - P_i)} - A_2 c_{d2i} \sqrt{\frac{2}{\rho}(P_i - P_{low})} \tag{5.13}$$

where P_p is the constant pump pressure, P_{low} is the constant reservoir pressure, A_1, A_2 are constants representing the orifice area, and ρ is the density of the fluid. P_i is the hydraulic pressure at the valves from the ith wheel cylinder side. Neglecting the inertial properties of the fluid and the resistance in the hydraulic line, it is assumed that the pressure in the wheel cylinder is also equal to P_i. The coefficients c_{d1i}, c_{d2i} are in fact the control inputs, which can take the values of 0 or 1 depending on the corresponding valve being open or closed. If the nonlinearities and the temperature dependence are neglected, the brake torque T_{bi} is a linear function of the brake pressure P_i

$$T_i = \left(P_i - P_{out}\right) A_{wc} \eta B_F r_r \tag{5.14}$$

where P_{out} is a push out pressure; A_{wc}, η, B_F, r_r are constants (wheel cylinder area, mechanical efficiency, brake factor, and effective rotor radius, respectively).

The rotational dynamics of the ith wheel ($i = 1, 2, 3, 4$) are modeled,

$$J\dot{w}_i = -T_{bi}sign\left(w_i\right) - R_e F_i + T_{di} \tag{5.15}$$

where

J: the moment of inertia of the wheel about the axis of rotation;

w_i: the angular velocity of the wheel;

T_{bi}: the brake torque at the ith wheel;

$R_e F_i$: the tire/road torque produced by the friction reaction force;

T_{di}: the engine torque, which is assumed to be zero during braking.

The linear dynamics of the vehicle is described as in simple cruise control system (see Section 5.1.2). According to the Newtonian equation,

$$M\dot{u} = \sum_{i=1}^{4} F_i - F_{load}$$

and the whole ABS system is constituted by the ninth-order nonlinear system involving four wheel dynamics, vehicle dynamics, and hydraulic actuator dynamics;

$$J\dot{w}_t = -T_{bi}sign\left(w_i\right) - R_e F_i + T_{di} \tag{5.16}$$

$$M\dot{u} = \sum_{i=1}^{4} F_i - F_{load} \tag{5.17}$$

$$c_f \frac{dP_i}{dt} = A_1 c_{d1i} \sqrt{\frac{2}{\rho}\left(P_p - P_i\right)} - A_2 c_{d2i} \sqrt{\frac{2}{\rho}\left(P_i - P_{low}\right)} \tag{5.18}$$

where $i = 1, 2, 3, 4$. There are eight control inputs c_{d1i}, c_{d2i}, $i = 1, 2, 3, 4$, which can take 0 or 1 with constraints c_{d1i}, $c_{d2i} = 0$.

The ABS control may be stated as follows: steering the slip at each wheel k_i to its extremal value $k_{*i}(t)$, and tracking this extremum during the braking transient.

The main difficulties in applying well-known linear control methods for this problem are:

- The system is strongly nonlinear.
- The control inputs can take only a finite number of values.

- The functions $F_i(t,k_i)$ are not a priori known and are, in general, nonstationary.
- Various parametric uncertainties and disturbances affect the ABS system.

5.3   Steering Control and Lane Following

5.3.1   Background

One of the key goals of an automated vehicle is the ability to perform automatic steering control. Steering control is a nontrivial design problem and a fundamental design challenge, and the approaches taken to obtain a stabilizing robust controller design vary significantly based on the available set of sensors and the performance of the actuators involved. A measure of the vehicle's orientation and position with respect to the road must be available to the controller. Among the most commonly used techniques are vision-based lane marker detection (preferred by many because of its simplicity in terms of the required machinery and implementation convenience), radar-based offset signal measurement (developed and used by OSU researchers exclusively), and the magnetic nail-based local position sensing (used by PATH researchers). Vision- and radar-based systems provide an offset signal at a preview distance ahead of the vehicle that contains relative orientation information. The vision system directly processes the image of the road and detects lane markers. Therefore, it does not require any modifications to current highway infrastructures. The radar system requires that an inexpensive passive frequency-selective stripe (FSS) be installed in the middle of the lane, in which case the radar is capable of providing preview information similar to a vision system. Most other sensor technologies provide only local orientation and position information. It has been pointed out that control of vehicles without preview distance measurements poses a difficult control problem at high speeds. Indeed, the experience of researchers using look-down-only sensors is that road curvature information must be provided to the lateral controller, usually by encoding it in the sensor components installed on the road. Thus we see that sensors are an integral part of the design and that the performance of the sensor system directly impacts the closed-loop system stability and performance.

5.3.2   Steering Control

In this section, a "circular" look-ahead (CLA) steering control is proposed to improve the curve following accuracy while taking into account of the curvature radius of the lane to be followed. The steering controller assures an automated guided vehicle to follow a desired path. On the path, three waypoints in front of the car, point A, B, and C are generated; they are illustrated in Figure 5.11. The distances between the waypoints and the distance from the vehicle center point to the first waypoint are L_{wp}, which depends on the vehicle speed V_s and the controller's update period T_{cycle} as given by (5.19). The three waypoint positions are sent to the path-following controller to provide path information.

$$L_{wp} = T_{cycle}V_s \qquad (5.19)$$

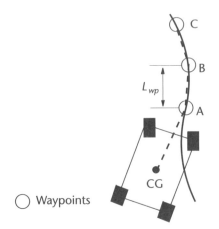

Figure 5.11   Sampling path.

Towards development of the lane-following method, a vehicle kinematic model, called the Dubin's car model, and the steering dynamics are presented. Under the assumption that there is no slip between the vehicle and road and the vehicle geometric center is at the center,

$$\dot{x} = V_s \cos\left(\theta_{yaw} + \frac{\theta_{steer}}{2}\right) \tag{5.20}$$

$$\dot{y} = V_s \sin\left(\theta_{yaw} + \frac{\theta_{steer}}{2}\right) \tag{5.21}$$

$$\omega = \frac{V_s}{R} = \frac{V_s}{L_{car}} \tan\theta_{steer} = \dot{\theta}_{yaw} \tag{5.22}$$

$$\theta_{SteerCom} = a\dot{\theta}_{steer} + b\theta_{steer} \tag{5.23}$$

where a and b are parameters of the steering and speed dynamics, θ_{steer} is the steering angle, $\theta_{SteerCom}$ is the steering command from the path following controller, V_s is the vehicle velocity, L_{car} is the car length, ω is the yaw rate, θ_{yaw} is the vehicle yaw angle, and R is the turning radius, as drawn in Figure 5.12.

The proposed controller is constituted by feedforward and feedback loops. Since the controller is implemented on a digital signal processor with discrete updating periods, the path curvature is predicted by the feedforward controller in the current and the coming update information period. The curvature may be obtained from the sampled waypoints in the path of the subject vehicle as illustrated in Figure 5.13. The feedforward control denoted by $\theta_{SteerFF}$ is derived by the sampled path points:

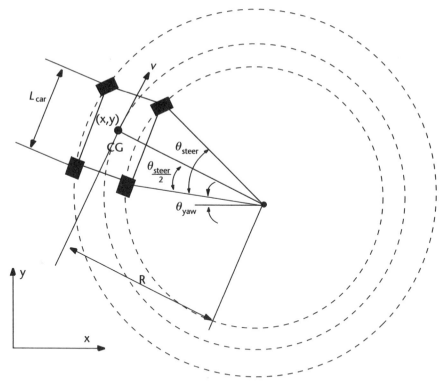

Figure 5.12   Vehicle model.

$$m_r = \frac{Y_B - Y_A}{X_B - X_A}$$

$$m_t = \frac{Y_C - Y_B}{X_C - X_B}$$

$$X_{ff} = \frac{m_r m_t (Y_C - Y_A) + m_r (X_B + X_C) - m_t (X_A + X_B)}{2(m_r - m_t)}$$

$$Y_{ff} = \frac{Y + Y_C}{2} - \frac{1}{m_t}\left(X_{ff} - \frac{X_B + X_C}{2}\right)$$

$$R_{ff} = \sqrt{(X_A - X_{ff})^2 + (Y_A - Y_{ff})^2}$$

$$\theta_{steer}FF = \tan^{-1}\left(\frac{L_{car}}{R_{ff}}\right)$$

(5.24)

The feedback scheme combines "circular look-ahead" (CLA) and a proportional, integral, and derivative (PID) controller to regulate the steering dynamics of the vehicle model. Unlike the general straight look-ahead, the CLA does not cut a curve, which is a reason to cause path follow error on a curve. As shown in Figure 5.14, for a straight look-ahead controller, the look-ahead offset D_{error}, the feedback to steering controller may not be generated even if a car is at a distance from the desired path. In this situation, the path following error is zero and the steering controller output does not change and it maintains the car exactly on the path. If

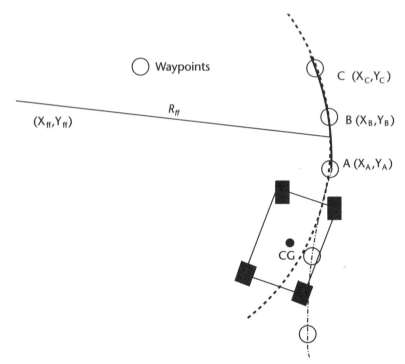

Figure 5.13 Feedforward control.

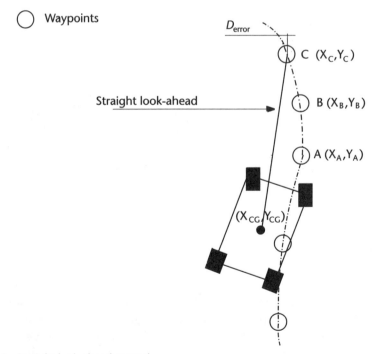

Figure 5.14 Straight look-ahead control.

there is no additional feedback for the steering control, a car keeps following a path with this offset.

The CLA predicts a future path based on the current steering angle instead of a straight look ahead. According to the current steering angle, a current circular path plotted with a solid line in Figure 5.15 is generated in ahead of the vehicle, which is a path it will go through if the steering angle is constant. Another circle, plotted with a dash-dot line, which is generated by the vehicle position and the two way-points ahead of the vehicle, points A and B, is the desired path that the car should follow. Comparing the two circles, the difference between the desired path and the current path can be found. The difference is defined by circles ahead of the offset. The ahead offset is the distance between the intersections of the two circles and a circle centered at the vehicle center with radius D_{la}, the circle plotted with a dashed line is situated at the center of the vehicle. D_{la} is also defined as the look-ahead distance. The ahead offset is the feedback of the steering control.

The equation of the desired path circle can be obtained from (5.25). The equation of the current path can be obtained as follows:

$$R_{cp} = \frac{L_{car}}{\tan\left(\theta_{steer}\right)}$$

$$X_{cp} = X_C - \frac{L_{car}}{2}\sin\left(\theta_{yaw}\right) - R_{cp}\cos\left(\theta_{yaw}\right)$$

$$Y_{cp} = Y_C - \frac{L_{car}}{2}\cos\left(\theta_{yaw}\right) + R_{cp}\sin\left(\theta_{yaw}\right) \tag{5.25}$$

$$\left(x - X_{cp}\right)^2 + \left(y - Y_{cp}\right)^2 = R_{cp}^{\ 2}$$

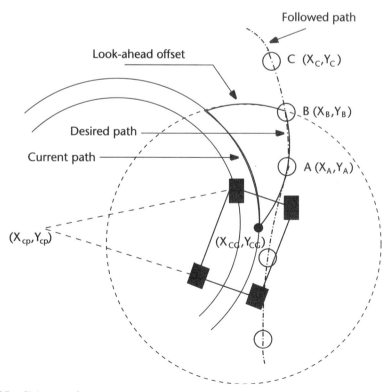

Figure 5.15 CLA control.

The ahead offset D_{os} with a look-ahead distance equal to D_{la} can be expressed as:

$$
\begin{aligned}
d1 &= \sqrt{\left(X_{cp} - X_c\right)^2 + \left(Y_{cp} - Y_c\right)^2} \\
a1 &= \left(R_{cp}^2 - D_{la}^2 + d1^2\right)/2d1 \\
h1 &= R_{cp}^2 - a1^2 \\
X1 &= X_c + a1\left(X_{cp} - X_c\right)/d1 \pm h1\left(Y_{cp} - Y_c\right)/d1 \\
Y1 &= Y_c + a1\left(Y_{cp} - Y_c\right)/d1 \mp h1\left(X_{cp} - X_c\right)/d1 \\
d2 &= \sqrt{\left(X_{dp} - X_c\right)^2 + \left(Y_{dp} - Y_c\right)^2} \\
a2 &= \left(R_{dp}^2 - D_{la}^2 + d2^2\right)/2d2 \\
h2 &= R_{dp}^2 - a2^2 \\
X2 &= X_c + a2\left(X_{dp} - X_c\right)/d2 \pm h2\left(Y_{dp} - Y_c\right)/d2 \\
Y2 &= Y_c + a2\left(Y_{dp} - Y\right)/d2 \mp h2\left(X_{dp} - X_c\right)/d2 \\
D_{os} &= \sqrt{\left(X2 - X1\right)^2 + \left(Y2 - Y1\right)^2}
\end{aligned}
\tag{5.26}
$$

The feedback part of the steering control is a PID controller, which can be expressed as:

$$
\theta_{steerFB} = K_1 D_{os} + K_2 \dot{D}_{os} + K_3 \int D_{os} dt
$$

The complete steering control will be:

$$
\theta_{SteerCom} = a\dot{\theta}_{steer} + b\theta_{steer}
$$

where K_1, K_2, and K_3 are weighted parameters.

In order to assure maneuvering task on the desired curvature or corner, the speed of the vehicle model may be regulated as an additional degree of freedom. The speed controller generates the speed reference and tracks this reference, which is the highest speed possible for a car to accurately follow a path. The speed reference is computed under the steering dynamics' time constant constraint, which is basically the amount of time needed to steer a steering wheel from the current angle to the next angle. The time needed to turn the steering mechanism should be smaller than the update period denoted by T_{cycle}. According to the update period and the difference between the current steering angle and the next steering angle, the desired speed is calculated.

The desired speed depends both on the current and the predicted next steering, the feedforward angle, and the time responses of the steering command are computed as:

$$
\theta_{steer}(t) = \frac{1}{b}\left(1 - e^{-\frac{b}{a}t}\right)\left(\theta_{SteerCom}(t + \Delta t) - \theta_{SteerCom}(t)\right)
\tag{5.27}
$$

For situations when the steering angle change $(\theta_{SteerCom}(t + \Delta_t) - \theta_{SteerCom}(t))$ is small, the rising time, the time needed to steer to $\theta_{SteerCom}(t + \Delta_t)$, is almost a constant $\dfrac{4.4a}{b}$. In such a situation the steering system should always be able to turn the steering wheel to the desired angle within an update period if $\dfrac{4.4a}{b} < T_{cycle}$.

However, the steering motor has finite output power to actuate the steering mechanism. When the steering angle change is larger than the threshold $\dfrac{4.4a}{b}(\dot{\theta}_{steer})_{max}$, where $(\dot{\theta}_{steer})_{max}$ is the maximum steering speed, the rising time becomes angle dependent, which can be approximated as:

$$T_{NextAngle} = \frac{\left(\theta_{SteerCom}(t + \Delta t) - \theta_{SteerCom}(t)\right)}{\left(\dot{\theta}_{steer}\right)_{max}} \tag{5.28}$$

In this situation the steering system may not be able to turn the steering wheel to the desired angle before the next update if the speed is too high. To make sure the steering angle can achieve to $\theta_{SteerCom}(t + \Delta_t)$ at the next update, the time needed to reach the next vehicle position, where the controller will be updated based on the current speed, should be greater than $T_{NextAngle}$. The distance to the next update point is $V_s\Delta t$, where V_s is the current speed and Δt is the update period equal to T_{cycle}. The maximum speed can be obtained:

$$v_{max} = \frac{\left(\dot{\theta}_{steer}\right)_{max} V_s \Delta t}{\theta_{SteerCom}(t + \Delta t) - \theta_{SteerCom}(t)} \tag{5.29}$$

The upperbound of the speed depends on the current steering angle, the predicted next steering angle, the steering dynamics' time response, and the actual velocity of the vehicle.

The path following algorithm is applied to a wheeled robot to drive in a simulated urban environment as shown in Figure 5.16. The difference between the robot and a vehicle is the robot uses differential drive for turning instead of steering wheels. The wheeled robot is a "Pioneer 3." Its location is determined by a camera on the ceiling based on its "tag" [see Figure 5.16(a)]. The location is forwarded like a GPS signal and sent to the robot by wireless.

To simulate a robot as a vehicle with steering wheels, the following equation is used to transform the steering angle θ_{steer} to the robot turn rate $\dot{\theta}_{YawRobot}$.

$$\dot{\theta}_{YawRobot} = \frac{V_s \tan\left(\theta_{steer}\right)}{L_{car}} \tag{5.30}$$

Figure 5.17 presents the experimental results from the wheeled robot. The circles are the assigned checkpoints the robot should go through and the lines are the robot path. The robot starts from the upper right of the figure and ends at the lower right. There are five curves in the path. The first two and the last two curves

(a)

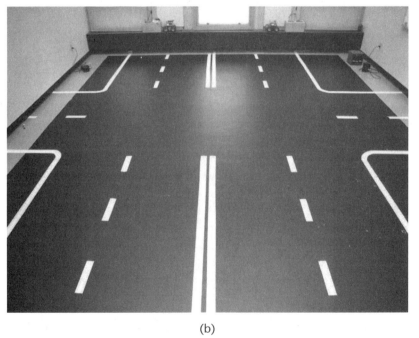

(b)

Figure 5.16   (a) The wheeled robot is regulated by CLA. (b) Simulated urban environment to perform steering control.

are sharp curves, which require fast steering, and the third curve is a long curve, which is used to verify the path following accuracy on a curve. Both curves have constant curvatures.

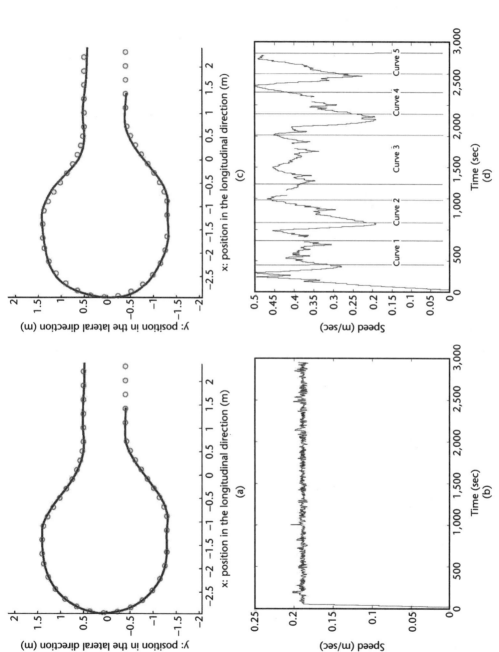

Figure 5.17   The time responses of robot trajectories and speed. For the case when speed is low, robot trajectories and speed responses are plotted in (a) and (b), and when speed is set to its maximum, robot trajectories and speed are plotted in (c) and (d).

During cornering maneuvering, path following error does not increase and the objective of the presented CLA steering control is validated experimentally. In the experimental scenario where the speed is low and kept constant, the robot can follow the path accurately [see Figure 5.17(a, b)]. Figure 5.17(c, d) represents the experimental results for the case when the speed is set to its maximum. The maximum wheel speed of the robot is 6 rad/sec, which corresponds to the maximum robot speed of 0.6 m/sec. Due to the differential turning employed by the robot, when the robot speed is high, the turning speeds of the outer turning wheels may require a speed higher than 6 rad/sec while turning. When this phenomenon occurs, the driving speed is reduced to assure each wheel rotates lower than 6 rad/sec.

Other than steering control there are more general obstacle avoidance approaches. The potential field approach is an example for obstacle avoidance methodology [25]. The idea is to have an attractive potential field function to get the car trajectory towards the waypoints to be followed and have an increasing repulsive potential function to push the trajectory from the obstacles. Using attractive and repulsive force analogy, the elastic band method is used to modify or deform the car trajectory locally [26]. An elastic band or bubble band is created such that the car can travel without collision. Once the obstacle is sensed, the elastic band is smoothly deformed to avoid obstacles. The external repulsive and internal attractive forces are generated to change the trajectory locally to avoid the obstacles and to keep the bubbles together, respectively. The potential field and elastic band obstacle avoidance approaches are illustrated in Figure 5.18(a, b), respectively.

The car may navigate autonomously in an environment where the obstacles may be nonstationary and whose locations may be unknown a priori. Therefore there are various instances where the obstacles need to be checked to avoid possible collision scenarios on the generated global path, which is started by the initial point and ended by the desired fixed final point. In general a global motion planning needs to be combined with a reactive strategy to continuously plan a motion plan while avoiding the obstacles. For example, a sampling-based approach is performed prior to computation by using a multiquery for the global motion planning, and a single-query–based computation on the fly for real-time obstacle avoidance. The reactive strategy, which computes obstacle location on a real-time basis, relies heavily on the obstacle information provided by the sensor fusion module. The obstacles are clustered and tracked by fusing the data collected from all sensors. A collision-free gap between the vehicle location and the intermediate waypoint towards the final point may be obtained by scanning the distance between the boundary of the tracked obstacles and the vehicle position. The free gaps are calculated by detecting the discontinuities in the scannings. In the sensing range, the obstacles are respresented by the disctontinuities and the free gaps are defined by the closed continuous intervals. The intervals are then compared with the vehicle physical size in order to check the safe passing stage in the candidate free gap. If a candidate interval connecting the vehicle location and the next local waypoint is computed, the car is driven towards the free gap, and the scanning and single-query–based computation is iteratively continued untill reaching the goal point [27]. The multiquery for global motion planning and a single-query–based computation on the fly are illustrated in Figure 5.18(c). Free-gap calculation is shown in Figure 5.18(d).

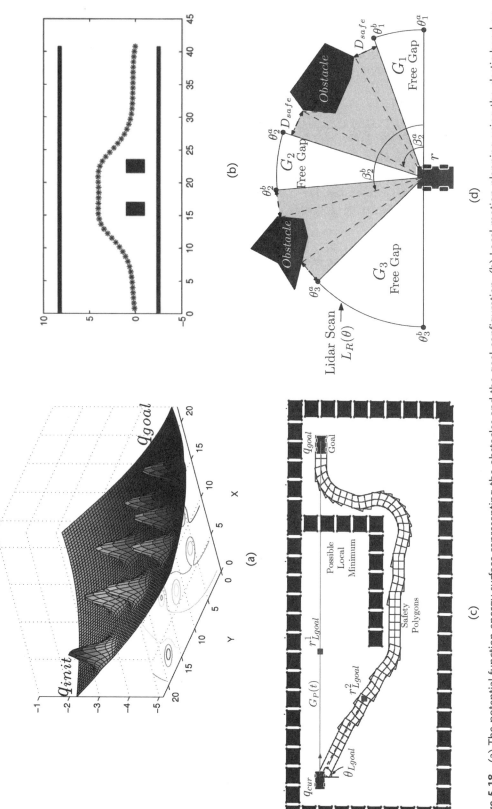

Figure 5.18 (a) The potential function energy surface representing the obstacles and the goal configuration. (b) Local motion planning using the elastic band approach. (c) Possible local goal selection using two different methods: the local goal r^1_{Lgoal} selected using the global path and r^2_{Lgoal} selected using the sampling-based path approach. (d) Illustration of the calculated free gaps by using a Lidar scan. A free gap is defined by its end angles.

5.3.3 Lane Following

It is assumed that the vehicle is operating on a flat surface and that a linearized bicycle model is capable of describing the motion of a vehicle effectively. (The bicycle model derivation is provided in the appendix.) The standard linearizing small-angle assumptions are made for the tire slip angles and the front tire steering angle. A wind disturbance is modeled that affects the lateral and yaw motions of the vehicle. The corresponding model is depicted in Figure 5.19. The variables represent the following physical quantities: $u(t)$, $v(t)$, and $r(t)$ are the longitudinal velocity, lateral velocity, and yaw rate, respectively; $\delta(t)$ is the actual steering angle of the front tires; $\psi(t)$ is the yaw angle with respect to the road; $y_{cg}(t)$ is the deviation of the vehicle's center of gravity from the lane center, $o(t)$ is the offset signal at the look-ahead point; F_{yf} and F_{yr} are the lateral tire forces on the front and rear tires, respectively; a and b are the distances from the center of gravity of the vehicle to the front and rear axles, respectively; I_w is the position at which a wind disturbance force of f_w laterally affects the vehicle motion; d is the sensor preview distance; and $\rho(t)$ is the road curvature at the look-ahead point. All distance measurements are in meters, and all angles are in radians.

The vehicle dynamics are represented by the following set of linear system equations

$$\dot{v}(t) = a_{11}v(t) + a_{12}r(t) + b_1\delta(t) + d_1 f_w$$

$$\dot{r}(t) = a_{21}v(t) + a_{22}r(t) + b_2\delta(t) + d_2 f_w$$

$$\dot{y}_{cg}(t) = v(t) + u\psi(t)$$

$$\dot{\psi}(t) = u\rho(t - t_0) - r(t) \tag{5.31}$$

$$\ddot{z}(t) = u^2\left[\rho(t) - \rho(t - t_0)\right] - du\,\dot{\rho}(t - t_0)$$

$$o(t) = y_{cg}(t) + d\psi(t) + z(t)$$

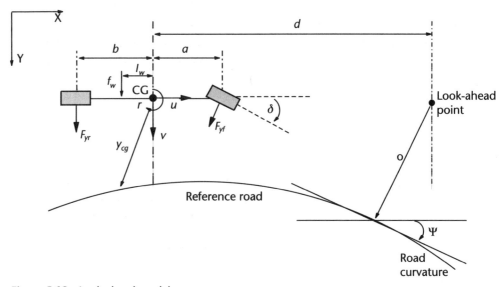

Figure 5.19 Look-ahead model.

where $z(t)$ is a dummy variable that is necessary to characterize the transient response of the offset signal precisely. In this context, $o(t)$ is the measured offset from the lane center at the look-ahead point (positive to the left of the lane center) and is to be regulated to zero for all possible road curvature reference inputs $\rho(t)$ (positive for clockwise turns); this defines the desired path to be followed using the front-wheel steering angle $\delta(t)$ (positive for clockwise turns). The linearized model is valid at the operating longitudinal velocity u (positive for forward motion), which is assumed to be kept constant by means of a decoupled longitudinal controller, and for small values of $\rho(t)$. The sensor delay t_o depends on the operating velocity and the preview distance d, and is given by $t_o = d/u$.

The other parameters of the vehicle model are determined from

$$a_{11} = -\frac{k_f + k_r}{mu}, \; a_{12} = -\left(u + \frac{ak_f - bk_r}{mu}\right), \; b_1 = \frac{k_f}{m}, \; d_1 = \frac{1}{m}$$

$$a_{21} = -\frac{ak_f - bk_r}{I_z u}, \; a_{22} = -\frac{a^2 k_f + b^2 k_r}{I_z u}, \; b_2 = \frac{ak_f}{I_z}, \; d_2 = -\frac{I_w}{I_z}$$

where $k_f > 0$ and $k_r > 0$ are the lateral tire stiffness coefficients of the front and rear tires, respectively, m is the mass of the vehicle, and I_z is the moment of inertia around the center of mass perpendicular to the plane in which the vehicle is located. The remaining variables are as previously defined. Typical parameter values approximating those of the OSU vehicles are given in Table 5.1.

5.3.3.1   A Nonlinear Lane-Keeping Controller

The vehicle model provided in the previous section is fairly generic. Here, on the other hand, we provide a specific controller (that used by the OSU team during Demo '97) simply to illustrate the type of nonlinear controller design that may be needed in an IV application.

The lateral control law that is employed to steer the vehicle, which consists of multiple terms that are functions of the measured signals, is

Table 5.1   Typical Model Parameters for OSU Vehicles

a	1.35m	CG to front axle distance
b	1.37m	CG to rear axle distance
m	1,569 kg	Total mass of the vehicle
k_f	5.96×10^4 N/rad	Front tire cornering stiffness
k_r	8.66×10^4 N/rad	Rear tire cornering stiffness
I_z	272.4 Ns/rad	Moment of inertia along z-axis
u	[1, 40] m/s	Range of longitudinal velocity
d	8.1m	Preview distance

$$\delta_{buf}(t) = K_d . \hat{\dot{o}}(t) + K_s . \hat{o}(t) |\hat{o}(t)| + K_\psi . \psi_{reset}(t)$$

$$+ K_r(r(t) - r_{ref}(t)) + K_i . sat\left(\int_0^t \hat{o}(\tau)d\tau\right) \qquad (5.32)$$

$$+ K_m . |p| . sign(deadzone(o(t)))$$

$$\delta_{com}(t) = sat(\delta_{buf}(t))$$

where $\hat{o}$ and $\hat{\dot{o}}$ are the Kalman observer estimates of the offset signal and its derivative and K_d, K_s, K_ψ, K_r, K_i, and K_m are gains of appropriate dimensions and signs; $\psi_{reset}(t)$ is defined as:

$$\psi_{reset}(t) = \int_0^t r(\tau)d\tau$$

such that $\psi_{reset}(t) = 0$, $\forall t = 0.5k$, $k \in N \cup \{0\}$.

Each component of the nonlinear steering signal given in (5.31) has a particular significance. The derivative of the offset signal helps suppress the otherwise noticeable limit cycles. The quadratic term generates a large penalty for large deviations from the lane center. The resetting yaw angle periodically corrects the orientation of the vehicle and aligns it with the road. The integral term is used to minimize the tracking error at steady state, and the saturation helps reduce oscillatory behavior during transients. The last term accounts for a sliding-mode–like switching assist toward the lane center upon necessity. A crossover detection algorithm along with a resettable timer runs continuously. If the vehicle deviates from the lane center for more than a specified time period and if its peak deviation, p, exceeds a threshold value, than an additive steering term nudges the vehicle toward the lane center until a crossover occurs. Under normal operating conditions where the vehicle is tracking the road center closely, this term has no contribution. During normal driving, the nominal component in the steering command is the term based on yaw error (the difference between a reference and the actual yaw rates). The overall steering command to the steering motor is saturated in order to satisfy safety requirements.

The parameters used for normal highway driving are shown in Table 5.2. A different set of parameter values are required for high-performance (high-speed, large curvature, winding, or slalom course) driving.

Table 5.2 Lateral Controller Parameter Values

K_d	12.00	K_s	46.00	K_ψ	−10.00
K_r	−1,200.00	K_i	12.00	K_m	75.00
K_{ref}	0.03	K_δ	−25.00	κ	2.00
T	0.01	B_1	1.00	B_2	100.00
P	0.00437	μ	1.00	γ	0.10
M_u	1.50	Δ_1	1.00	Δ_2	1.00

5.3.3.2   A Lateral Lane Change Controller

In any application-oriented controller design, the reasoning behind the design path pursued lies in the plant to be controlled and the available sensors and actuators. Modularity and flexibility are always desirable, but the controller must work on the system at hand. In this case, the choice of a design procedure was mandated by the fact that preview sensor information (which is used in our lane-keeping algorithm) cannot be measured continuously during the transition from one lane to another using either the vision or the radar reflective sensors. There is a dead-zone period when the preview sensing systems do not provide useful data. This creates a transition period that must be handled "open-loop" with respect to lateral position information. Attempts to generate a true open-loop time series steering angle command profile failed because of wind and super-elevation disturbances, nonsmooth actuator nonlinearities, unmodeled vehicle-road interactions, and uncertainties in the (possibly time-varying) plant parameters. Most of these dynamics and disturbances can be bypassed through the yaw rate measurement. Thus for the lane change a vehicle yaw rate controller was designed and used to implement a desired time series yaw rate profile, which would bring the vehicle to the center of the next lane and preserve the vehicle's angular alignment with the road.

The lane change problem can be summarized as follows: While maintaining lane orientation at a longitudinal speed u, the vehicle travels a specified distance (a full lane width) along the lateral axis with respect to its body orientation within a finite time period and aligns itself with the adjacent lane at the end of the maneuver such that the lane-keeping task can be resumed safely and smoothly. The autonomous lane change problem deals with the generation of the appropriate steering signal to cause the vehicle to accomplish the above-described task without driver assistance. The major design assumptions are:

1. Only the yaw rate r and the steering angle δ are measured.
2. Vehicle parameters are known within a bounded neighborhood of some nominal values.
3. The road curvature does not change significantly during the lane change maneuver.

Studies have been performed to estimate the ideal lateral jerk, acceleration, velocity, and displacement signals that the vehicle's center of gravity should follow to perform a lane change maneuver while preserving passenger comfort. However, in practice the only input to the vehicle is commanded steering angle. Therefore, these results must ultimately be used to generate steering angle commands. This can be accomplished by generating a reference yaw rate signal and applying a yaw rate controller to generate steering angle commands.

Through this steering control subsection, the operating region in the tire lateral characteristics is assumed to be responsive to the steering inputs and handling in the lateral maneuvering tasks is performed. Although the nonlinear tire characteristics and handling properties are out of scope of this book, the interested reader is advised to investigate handling and vehicle dynamics control issues. An alternative semiempirical tire model and vehicle dynamics control issues are given in [28, 29].

One of the factors affecting the motion and handling dynamics is obviously the vehicle speed. During maneuvering, lateral tire force responses with respect to the slip ratio and the sideslip angle are plotted in Figure 5.20. In the normal operating region (i.e., at low values of the slip in the longitudinal direction), lateral force generation is responsive to the sideslip angle increment. Estimation of the slip ratio may be very complicated due to the unknown road-tire friction coefficient. Therefore, active safety enforces maintaining a safe speed to limit sideslip and possibly prevent rollover hazards that may occur during maneuvering at high speed. Decelerating to a lower speed before a sharp turn is a common driver's instinct in order to maintain control authority during a lateral maneuver. A study on rollover prevention for heavy trucks carrying liquid cargo is presented in [30]. Differential braking modulated by the nonlinear control techniques is used to stabilize the lateral dynamics.

5.4 Parking

Parking is basically displacing the vehicle to the final position from any arbitrary initial condition. Parking process establishes path planning for the vehicle at low speed with a desired heading angle while avoiding any possible obstacles situated near the parking lot. From the viewpoint of motion planning, the parking path is a circular arc to be tracked by the parking algorithm. At the initial parking local setting, initial yaw angle of the car model may be different from the tangent angle of the path. To achieve parking, the tangent angle of the path needs to be reached. Driving forward or backward by turning the front steering wheel to its maximum

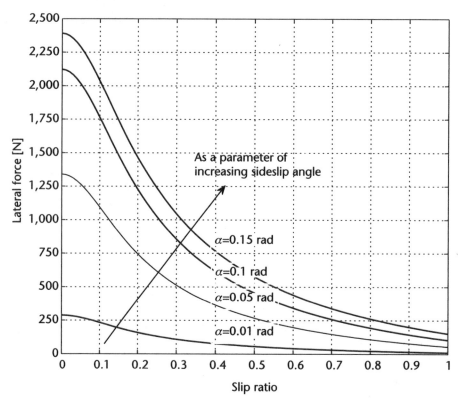

Figure 5.20   Lateral tire characteristics at different slip angles.

allowable angle may help to refer the vehicle heading angle to the desired parking path tangent angle [31]. Once the tangent of the path is reached, the circular arc is tracked until the desired parking position. The vehicle model with Ackerman steering (Figure 5.21) is used,

$$\theta_{steer} = \frac{\theta_l + \theta_r}{2}$$
$$R = \frac{L_{fr}}{\tan \theta_{steer}}$$

(5.33)

where R is the vehicle turning radius, θ_{steer} is the average steering angle, and θ_l and θ_r denote the steering angle of the front left wheel and front right wheel, respectively. Following the Ackermann steering theory, the vehicle model position is equal to the center of the rear axle and the yaw angle is regulated to follow the arc path direction.

5.4.1   Local Coordinates

The origin of the local coordinate is defined to be the center of the parking area in the horizontal direction whereas the origin should be as close to the final position as possible in the vertical direction; the car model is steered to reach this point under

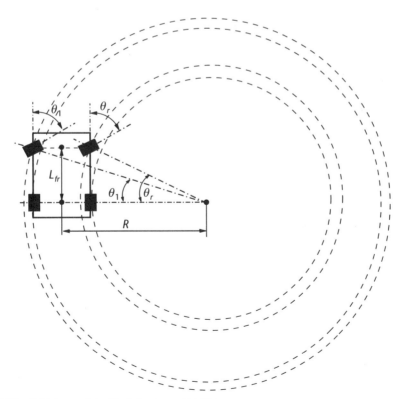

Figure 5.21   Vehicle model with Ackerman steering.

constraints of maximum steering turn radius and without hitting any boundaries or obstacles.

$$R_{min} = R_{turn} - W_c / 2$$

$$R_{max} = \begin{cases} \sqrt{(R_{turn} + W_c / 2)^2 + L_{fr}^2} & Parking \quad Forward \\ \sqrt{(R_{turn} + W_c / 2)^2 + (L_c - L_{fr})^2} & Parking \quad Backward \end{cases}$$

$$X0 = Wp / 2 \qquad\qquad\qquad\qquad\qquad\qquad\qquad\qquad\qquad\qquad\qquad (5.34)$$

$$Y0 = \begin{cases} \sqrt{R_{max}^2 - (R_{min} + (Wp / 2)^2)^2} & when \quad (R_{max}^2 - (R_{min} + (Wp / 2)^2)^2) >= 0 \\ -\sqrt{(W_c Wp / 2) - (W_c / 4)^2} & when \quad (R_{max}^2 - (R_{min} + (Wp / 2)^2)^2) < 0 \end{cases}$$

where Wp is the width of the parking spot, Wc is the car width, L_c is the car length, L_{fr} is the distance from front bumper to rear axle, R_{turn} is the vehicle minimum turning radius by the vehicle center, and R_{min}/R_{max} are the radiuses of the minimum/maximum circles a vehicle body can go through according to R_{turn}. The minimum turning radius of the car model may be obtained from the vehicle specification.

5.4.2 Parking Scenarios: General Parking Scenario and DARPA Urban Challenge Autonomous Vehicle Parking Scenario

The parking path is originally a circular arc to be tracked by the parking algorithm. At the initial parking local setting, initial yaw angle of the car model may be different from the tangent angle of the path as illustrated by Figure 5.22. To reach the tangent angle of the path, driving forward or backward by turning the front steering wheel to its maximum allowable angle is considered. Once the tangent of the path is reached, the circular arc is tracked until the desired parking position, as illustrated in Figure 5.23. A general parking scenario where the car may pull either forward or backward into the parking spot is presented. Three parking steps are illustrated in Figure 5.24.

In the DARPA Urban Challenge competition, a parking lot is defined as a "zone." The OSU-ACT (see, for example, Figure 1.4) is controlled by an "obstacle avoidance controller" in the zones. The obstacle avoidance controller can keep the car away from obstacles and bring it to a desired point. For the parking case:

- The goal point of the obstacle avoidance controller is the final position of the parking spot.
- The obstacle avoidance controller hands over the control authorities to "parking" when the distance to the final point is smaller than 8 meters.
- In the 2007 Urban Challenge, DARPA claimed that paths to a parking spot would not be totally blocked. This maneuver can make sure parking starts from the side that is obstacle free.

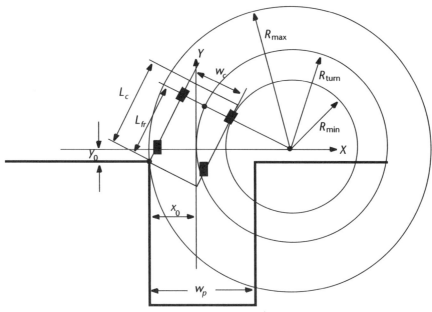

Figure 5.22   Local coordinate.

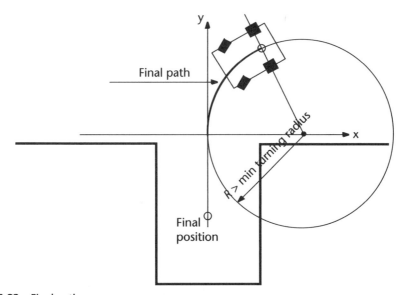

Figure 5.23   Final path.

- Pulling a car forward into the parking spot was imposed by a rule in the DARPA Urban Challenge.

The following strategies were obeyed to execute autonomous parking: Initially, the car goes forward to approach the parking spot. In order to decrease the rear steering wheel angle, denoted by θ_r, as illustrated in Figure 5.24. The steering direction [Figure 5.24(a)] θ_{steer} is derived by the following equation set

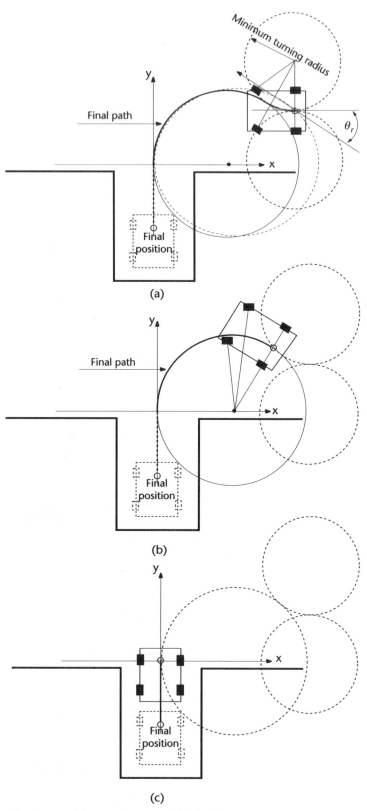

Figure 5.24   (a) First step, (b) second step, and (c) third step.

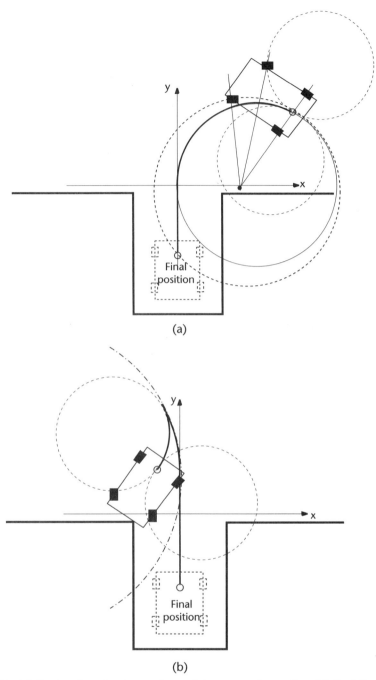

Figure 5.25 (a) Final path radius larger than minimum turning radius. (b) Find final path from second maneuver.

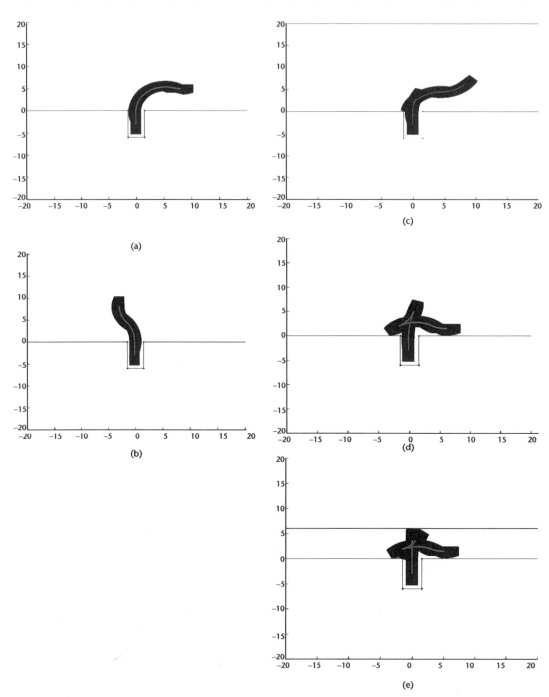

Figure 5.26 (a) General situation. (b) Park from different initial position. (c) Initial position close to boundary. (d) Initial position close to boundary. (e) Park at narrow road.

$$R = \frac{x^2 + y^2}{2|x|}$$

$$\theta_{FinalPath} = -\sin^{-1}\left(\frac{y}{R}\right) + \frac{\pi}{2}$$

$$\theta_{FinalPath} - \theta_{yaw} = \theta_r$$

$$\theta_{steer} = \begin{cases} left & \text{if} & (\theta_{FinalPath} > \theta_{yaw}) \\ right & \text{if} & (\theta_{FinalPath} < \theta_{yaw}) \end{cases}$$

where R is the radius of the possible final path circle, the circle with a larger diameter in Figure 5.24(c). x and y denote the vehicle position according to the local coordinate, $\theta_{FinalPath}$ is the tangent angle of the possible final path circle at the vehicle position, and θ_{yaw} is the vehicle's yaw angle. The final path is achieved when $\theta_{FinalPath} = \theta_{yaw}$.

This maneuvering may definitely be ended by an achievement of the final path circle. In case the radius of the final path circle is smaller than the minimum turning radius (i.e., path circle) which may not be followed by the vehicle model due to its turning limitation, as illustrated in Figure 5.25(a). The autonomous car keeps on going forward to minimize θ_r. The car will follow the circle plotted with the dash line in Figure 5.25(a). During this course, an auxiliary circle, plotted with a dash line, may be generated towards a final path circle (plotted with a dash-dot line) with a larger radius than the minimum turning radius as shown in Figure 5.25(b).

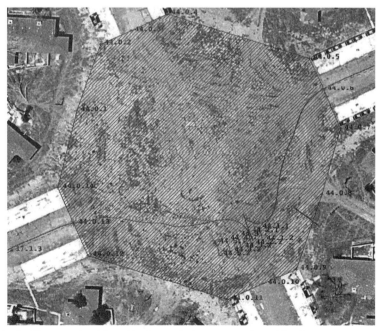

Figure 5.27   Parking area in Urban Challenge 2007. Approach and backup.

5.4.3 Simulation and Experimental Results

Simulation results are presented to illustrate the effectiveness of the considered parking algorithm. In Figure 5.26, simulations with different initial conditions are illustrated. The road is narrow, which makes parking complicated and a difficult task for the human being driver. The displacement of the car model with respect to a parking lot shows that parking is executed successfully after two forward-backward maneuverings.

The parking algorithm is applied successfully on the autonomous vehicle OSU-ACT on a real-time basis. In Figure 5.27 a satellite view of the zone in the DARPA Urban Challenge is given. The line is the path tracked by the OSU-ACT autonomous vehicle. The parking area can be seen by the parking motion trace (parking

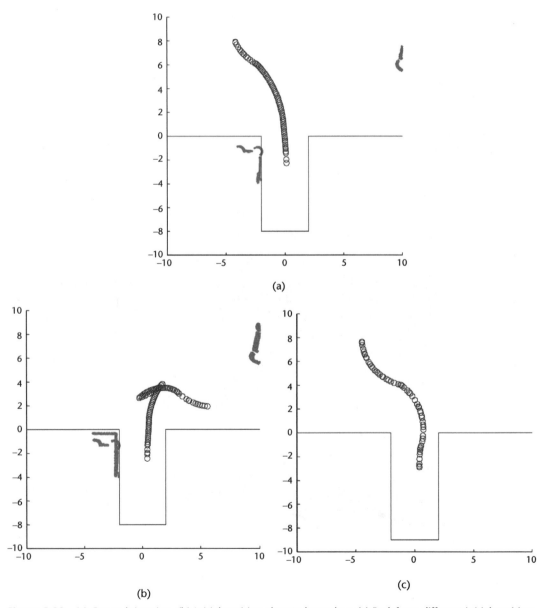

Figure 5.28 (a) General situation. (b) Initial position close to boundary. (c) Park from different initial position.

and backing up to be back on the path). In the parking area, two other cars were parked on the left and right sides of the assigned parking spot. The OSU-ACT was successfully parked into the parking spot (zone) at each round.

Experimental results with different initial conditions and stationary obstacle positioning are presented in Figure 5.28. The splines present the vehicle trajectory, solid lines are the parking zone boundaries, and the distributed points are obstacles detected by the LIDAR scanning.

References

[1] Hatipoglu, C., Ü. Özgüner, and M. Sommerville, "Longitudinal Headway Control of Autonomous Vehicles," *Proceedings of IEEE International Conference on Control Applications*, 1996, pp. 721–726.

[2] Ioannou, P. A., and Z. Xu, "Intelligent Cruise Control: Theory and Experiment," *Proceedings of the 32st IEEE Conference on Decision and Control*, 1993, pp. 1885–1890.

[3] Ioannou, P. A., and C. C. Chien, "Autonomous Intelligent Cruise Control," *IEEE Transactions on Vehicular Technology*, Vol. 42, 1993, pp. 657–672.

[4] Lu, X. Y., and J. K. Hedrick, "ACC/CACC Control Design, Stability and Robust Performance," *Proceedings of American Control Conference*, 2002, pp. 4327–4332.

[5] Lu, X. Y., H. S. Tan, and J. K. Hedrick, "Nonlinear Longitudinal Controller Implementation and Comparison for Automated Cars," *Journal of Dynamic Systems, Measurement and Control*, Vol. 123, 2001, pp. 161–167.

[6] Swaroop, D., et al., "A Comparison of Spacing and Headway Control Laws for Automatically Controller Vehicles," *Vehicle System Dynamics*, Vol. 23, 1994, pp. 597–625.

[7] Shladover, S. E., "Review of the State of Development of Advanced Vehicle Control Systems," *Vehicle System Dynamics*, Vol. 24, 1995, pp. 551–595.

[8] Tsugawa, S., et al., "An Architecture for Cooperative Driving of Automated Vehicles," *IEEE Intelligent Transportation Systems Conference Proceedings*, Dearborn, MI, 2000, pp. 422–427.

[9] Cho, D., and J. K. Hedrick, "Automotive Powertrain Modeling for Control," *Transactions of ASME*, Vol. 111, 1989, pp. 568–577.

[10] Kotwicki, A. J., "Dynamic Models for Torque Converter Equipped Vehicles," SAE paper no. 820393, 1982.

[11] Molineros, J., and R. Sharma, "Real-Time Tracking of Multiple Objects Using Fiducials for Augmented Reality," *Transactions on Real-Time Imaging*, Vol. 7, No. 6, 2001, pp. 495–506.

[12] Orqueda, A. A., and R. Fierro, "Visual Tracking of Mobile Robots in Formation," *Proceedings of the 2007 American Control Conference*, New York, 2007, pp. 5940–5945.

[13] Gonzales, J. P., "Computer Vision Tools for Intelligent Vehicles: Tag Identification, Distance and Offset Measurement, Lane and Obstacle Detection," Master Thesis, The Ohio State University, Electrical Engineering Dept., September 1999.

[14] Emig, R., H. Goebels, and H. J. Schramm, "Antilock Braking Systems (ABS) for Commercial Vehicles—Status 1990 and Future Prospects," *Proceedings of the International Congress on Transportation Electronics*, 1990, pp. 515–523.

[15] Tan, H. S., "Discrete-Time Controller Design for Robust Vehicle Traction," *IEEE Control Systems Magazine*, Vol. 10, No. 3, 1990, pp. 107–113.

[16] Johansen, T. A., et al., "Hybrid Control Strategies in ABS," *Proceedings of the American Control Conference*, 2001, pp. 1704–1705.

[17] Nouillant, C., et al., "Hybrid Control Architecture for Automobile ABS," *IEEE International Workshop on Robot and Human Interactive Communication*, 2001, pp. 493–498.

[18]   Johansen, T. A., et al., "Gain Scheduled Wheel Slip Control in Automotive Brake Systems," *IEEE Transactions on Control Systems Technology,* Vol. 11, No. 6, 2003, pp. 799–811.

[19]   Gustafsson, F., "Slip-Based Tire-Road Friction Estimation," *Automatica,* Vol. 33, No. 6, 1997, pp. 1087–1099.

[20]   Haskara, I., C. Hatipoglu, and Ü. Özgüner, "Sliding Mode Compensation, Estimation and Optimization Methods in Automotive Control," *Lecture Notes in Control and Information Sciences, Variable Structure Systems: Towards the 21st Century,* Vol. 274, 2002, pp. 155–174.

[21]   Drakunov, S., et al., "ABS Control Using Optimum Search Via Sliding Modes," *IEEE Transactions on Control Systems Technology,* Vol. 3, No. 1, 1995, pp. 79–85.

[22]   Yu, H., and Ü. Özgüner, "Extremum-Seeking Control Strategy for ABS System with Time Delay," *Proceedings of the American Control Conference,* 2002, pp. 3753–3758.

[23]   Yu, H., and Ü. Özgüner, "Smooth Extremum-Seeking Control Via Second Order Sliding Mode," *Proceedings of the American Control Conference,* 2003, pp. 3248–3253.

[24]   Bakker, E. T., H. B. Pacejka, and L. Linder, "A New Tire Model with an Application in Vehicle Dynamics Studies," SAE Technical Paper no. 870421, 1982.

[25]   Khatib, O., "Real-Time Obstacle Avoidance for Manipulators and Mobile Robots," *International Journal of Robotics Research,* Vol. 5, No. 1, 1986, pp. 90–98.

[26]   Quinlan, S., and O. Khatib, "Elastic Bands: Connecting Path Planning and Control," *Proceedings of Robotics and Automation,* Atlanta, GA, 1993, pp. 802–807.

[27]   Shah, A. B., "An Obstacle Avoidance Strategy for the 2007 Darpa Urban Challenge," Master's Thesis, The Ohio State University, Electrical Engineering Dept., June 2008.

[28]   Pacejka, H. B., *Tyre and Vehicle Dynamics*, Oxford, U.K.: Butterworth-Heinemann, 2002.

[29]   Rajamani, R., *Vehicle Dynamics and Control*, New York: Springer, 2006.

[30]   Acarman, T., and Ü. Özgüner, "Rollover Prevention for Heavy Trucks Using Frequency Shaped Sliding Mode Control," *Vehicle System Dynamics*, Vol. 44, No. 10, 2006, pp. 737–762.

[31]   Hsieh, M. F., and Ü. Özgüner, "A Parking Algorithm for an Autonomous Vehicle," *2008 IEEE Intelligent Vehicles Symposium*, Eindhoven, the Netherlands, 2008 pp. 1155–1160.

CHAPTER 6

Maps and Path Planning

Autonomy requires an understanding of the world. Local sensors can provide such information in a neighborhood of the vehicle of radius 50–100 meters. Perhaps cooperative vehicle-to-vehicle communication can extend that region to a radius of several hundred meters. In any case, the information available is of a local nature, suitable for short-term vehicle control but not for long-term planning. A fully autonomous vehicle will require planning behaviors over a much longer timescale. Ideally, a fully autonomous vehicle would allow navigation from any origin to any destination without direct human input.

This level of autonomy requires geographic information for an area and a planning algorithm that can use these maps to generate a plan for the vehicle path and behavior, as well as mechanisms for replanning when actual current conditions do not match the contents of the maps. The use of map databases is reasonable due to the availability of GPS-based positioning.

In this chapter, we will review the common types of map data available for off-road and on-road situations as well as path planning algorithms suitable for each situation.

6.1 Map Databases

Generally speaking, the available map data falls into two categories: raster data and vector data. Raster-based data divides an area into a collection of cells or grids, often of uniform size, and each grid cell is assigned one or more values representing the information being mapped. This approach is similar to the grid-based sensor fusion algorithms that were introduced in Chapter 4. Generally speaking, raster data consumes a large amount of memory but requires simpler data processing techniques as the information is represented at a more direct level. Examples of raster data include digital elevation maps and land cover or usage maps.

Vector data expresses information in terms of curves, connected line segments, or even discrete points. Curves or line segments can also enclose and represent areas. Vector data is a more complex and rich representation and thus generally requires less storage but more complicated processing algorithms. Examples include digital line graph road maps, constant altitude topographic maps, and real estate parcel databases.

Often, different types of data are geographically correlated and stored in a single map database. Each type of data is known as a "layer," as shown in Figure 6.1. For example, the pixel data from satellite images at different optical frequencies (infrared, visible green, visible red, and ultraviolet) might each be a layer.

6.1.1 Raster Map Data

In general terms, it is likely that information represented as raster data is useful for off road driving situations, though this is not universally true. A number of different sources, types of map data, and formats are available, including [1]:

1. Digital elevation map (DEM) data at 10-m resolution are widely available from the USGS NED database for most of the continental United States. The resolution may vary, for example in undeveloped desert areas the resolution may only be 30m. This provides only a very coarse terrain description, obviously not meeting the resolution and accuracy requirements of autonomous vehicle navigation. Furthermore, most of the USGS DEM data is rather old, which may pose a problem in some developed areas.
2. Space Shuttle radar topographic mission (SRTM) data is available but its resolution is still a modest 30m. However, as the data was taken in 2000, it is more recent than the NED datasets. The ASTER global digital elevation map dataset has also recently been released, although it too has a 30-m resolution.
3. USGS land cover data (metadata) is available for many areas.
4. USGS digital orthophoto quarter quadrangles (DOQQ) provide aerial image data at 1-m resolution.

Figure 6.2 is an example of data plotted from a U.S. Geological Survey base map called the 7.5-minute DEM, containing terrain elevation for ground positions at regularly sampled (approximately 30m) horizontal intervals referenced to the Universal Transverse Mercator (UTM) projection. This particular area is near Sturgis, South Dakota. These datasets are available online at http://edc.usgs.gov/products/elevation/dem.html and http://seamless.usgs.gov.

Land cover information can also be gathered from the National Landcover Characterization Dataset (NLCD). Aerial photograph and digital orthophoto quarter quadrangle (DOQQ) images are useful to visualize the path and verify its accuracy. A digital orthophoto is a digital image of an aerial photograph in which displacements caused by the camera and the terrain have been removed. It

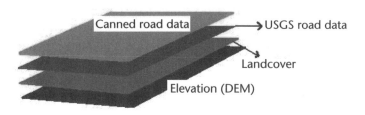

Figure 6.1 Layers in a GIS or map database.

Figure 6.2   7.5-minute DEM. (http://edc.usgs.gov/products/elevation/dem.html).

combines the image characteristics of a photograph with the geometric qualities of a map. The images are more useful than ordinary aerial photography because they are geo-referenced and ortho-rectified to remove camera and viewpoint perspective distortion. In addition to verifying the route they can be utilized as a useful tool for the revision of digital line graphs and topographic maps. Figure 6.3 shows a 1-m resolution orthophoto quarter quadrangle of the Washington, D.C., area.

6.1.2   Vector Map Data

In general terms, information represented as vector data is interpreted at a higher level and is useful for on-road driving situations. A number of different sources are available, some freely available and some commercial.

1. USGS digital line graphs (DLG) extracted from existing maps and other sources, which include roads, railroads, pipelines, constant elevation, hydrological features, political boundaries, and a variety of point features.

Figure 6.3   DOQQ of Washington, D.C. (http://edc.usgs.gov/products/aerial/doq.html).

2. TIGER/line (Topologically Integrated Geographic Encoding and Referencing System) data prepared by the U.S. Census Bureau, which provides more current information similar to the USGS maps and also street address data.
3. Commercial street map datasets, for example, DeLorme, Tele Atlas, Navtech.

It should be noted that, in general, currently available map data is neither sufficiently accurate for autonomous control purposes nor sufficiently detailed for complete behavior and path planning. For example, the streets may only be represented by a road centerline.

6.1.3   Utilizing the Map Data

Raster data can be processed in much the same way as grid cell sensor data for obstacle avoidance and planning. Possible algorithms include potential field methods, Voronoi decompositions with grid maps treated as graphs, and discrete gradient descent optimization approaches.

Vector data can be processed using tools from graph theory, a branch of discrete mathematics. A graph consists of nodes, also known as vertices, and edges. Edges may be directed and weighted. A road map can easily be translated into a graph by, for example, labeling intersections as nodes and streets as the edges connecting nodes. Various properties of a given graph, including connectedness, adjacency, reachability, and optimal flow and routing, can be computed from this representation.

The basic algorithm for optimal routing is known as Dijkstra's algorithm. When applied to a graph with non-negative edge path costs, it finds the lowest cost path between a given vertex and every other vertex in the graph. The algorithm is as follows:

1.   Given some initial node, mark its value as 0 and the value of all other nodes as infinite;
2.   Mark all nodes as unvisited.
3.   Mark the given initial node as the current node.
4.   For all unvisited neighbors of the current node:
     a.   Calculate a cost value as the sum of the value of the current node and the connecting edge.
     b.   If the current value of the node is greater than the newly calculated cost, set the value of the node equal to the newly calculated cost and store a pointer to the current node.
5.   Mark the current node as visited. Note that this finalizes the cost value of the current node.
6.   If any unvisited nodes remain, find the lowest cost unvisited node, set it as the current node, and continue from step 4.

If the user is only interested in the lowest cost path from the initial node to a given final node, the algorithm may be terminated once the final node is marked as visited. One can then backtrack through the node pointers to find the route corresponding to the lowest cost.

Figure 6.4 illustrates a simple execution of Dikstra's algorithm. Figure 6.4(a) gives a sample graph network, consisting of five nodes and connections with weights (or lengths) as shown. The origin is node 1 and the desired destination is node 5. As we begin the process [Figure 6.4(b)], the current node, shown with a double circle, is set to the initial node, its cost is set to 0, and the cost of all other nodes is set to infinity. In the first pass [Figure 6.4(c)], the cost for all nodes connected to the current node is set to the value of the current node, in this case 0, plus the edge cost and a pointer (the dashed line) is set to point back to the current node. In the second pass [Figure 6.4(d)], the current node is chosen as the lowest cost unvisited node, in this case node 4, and node 0 is marked as visited (crossed out in this example). The cost to node 3 is unchanged since the value of the current cost of node 4 plus the cost of the edge to node 3 is higher than the current cost of node 3. The cost for node 5 is set to the sum of the cost for node 4 and the cost of the edge between node 4 and node 5. Since node 5 is the destination, the algorithm would normally stop at this point, reporting the optimal value of the cost from node 1 to node 5 as 4 and the optimal path as $1 \rightarrow 4 \rightarrow 5$. Note however that the value of the current cost of node 2 is not the optimal value. If we wished to complete the calculation of the cost of node 2, we would need to take one more step, in which the current node would be switched to node 3, node 4 would be marked as visited, and the value of the cost of node 2 would become 5 instead of 6.

The Dijkstra algorithm is generally considered the basis for optimal path planning and routing algorithms. However, it is computationally expensive and so a number of extensions, usually employing some heuristic methodology to obtain an efficient though probably suboptimal solution, have been proposed, including the A* algorithm, which will be discussed in Section 6.2.

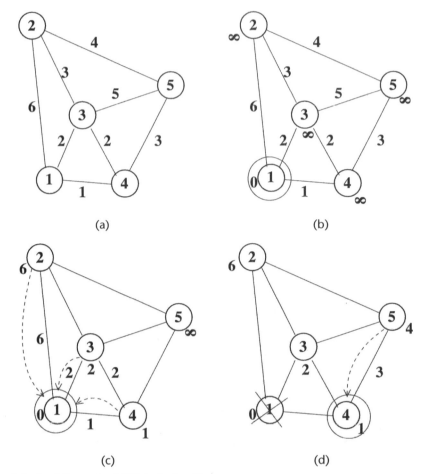

(a)   (b)

(c)   (d)

Figure 6.4   (a–d) Example of Dijkstra's algorithm.

6.2   Path Planning

The role of a path planning software module is to provide the coordinates of a traversable route connecting a given initial and desired final position. The following section discusses some of the issues for the architecture of a path planning module [1]. It also discusses the heuristics used in the implementation of path planning software.

Typically, the path planning task can be divided into two parts:

- *Global path planning:* This subtask deals with generating the best route to reach the destination in the form of a series of road segments or points such that the route is divided into dense evenly spaced segments.

- *Local path planning:* This subtask performs the path planning for each segment generated by the global path planner. Hence, the local path planner would be responsible for obstacle avoidance of unknown obstacles. It would need to be equipped with capabilities such as perceiving objects, road following, and maintaining continuous real-time motion.

Autonomous navigation is a complex problem and several features of the environment and the robot have to be considered before global path planning. The following features need to be taken into account in this case:

- *Incomplete knowledge:* In the real world the environment is not static, that is, the position of the obstacles is continuously changing. As a result, algorithms developed for path planning should accommodate these uncertainties. One approch is to use sensors to update the map as a change occurs in it or as problems with the currently planned route are identified. Another approach, suitable mainly for off-road navigation, is to employ an obstacle avoidance algorithm to circumvent obstacles that were not indicated on the map.

- *Robot dynamics:* The global path planner should consider the vehicle dynamics such as turning radius, dimensions of the vehicle, and maximum negotiable elevation so as to generate a safe and efficient route. Properties like inertia, response time to steering command, stopping distance, and maximum speed of motion are important for motion planning on the route generated by the global planner. These factors can be incorporated into the "cost function" of the algorithm.

- *Terrain dynamics:* It is important to know the behavior of the vehicle when operating on different terrain types. For example some vehicles may be not equipped to handle soft soil and could sink in it. Extreme terrain such as very rocky conditions could also pose a serious hazard, and a steep slope increases load on the drive train and also raises the center of gravity, which may destabilize the vehicle.

- *Global considerations:* Knowledge about the global conditions such as dangerous regions or resource limitations is yet another decisive factor for the global planner. It should compute the route in such a way that the robot achieves a near-continuous motion. The goal of the global planner is to ensure the most optimal path, that is, a path that satisfies at least some of the requirements like cost, energy, safety, or mechanical constraints for any given situation. The algorithm minimizes the cost function, thus optimizing the constraints mentioned above.

Considering the above factors, the implementation steps of the path planning module can be broadly classified into:

1. Preparing the map dataset or map database;
2. Road and/or terrain representation and interpretation from the map dataset;
3. Computing the optimal route using the mapping information.

6.2.1 Path Planning in an Off-Road Environment

As discussed earlier, path planning in outdoor terrain poses challenges because the environment is changing continuously. If the terrain is mapped accurately then the problem can be solved using conventional off-line path planning techniques. On the other hand, if there is no information available about the terrain then the vehicle has to rely on its sensors completely for gathering information about the surroundings. This would make the process too slow and computationally expensive since the sensor data has to be collected first and then processed before the global planner can use this information. Considering both of these alternatives, it can be concluded that some prior information about the surroundings, even if it is incomplete, is highly advantageous for the path planning module [2].

Terrain information is typically described by:

- Surface geometry or a geographical map;
- Thematic data or an object map.

Surface data is usually available in the form of an evenly spaced grid format such as the digital elevation model (DEM), in which only the ground surface is considered. Another topographic model of the Earth's surface is the digital surface model (DSM). This format depicts the top of all surfaces whether bare-earth, non-elevated manmade surfaces, elevated vegetative, or elevated manmade surfaces. It is often necessary to segment this type of data into smaller areas to reduce the computational processing and memory requirements.

The map databases give us information about elevation, roads, and land cover for the designated terrain. The path planning software has to access this database and use the information to optimize the route based on a prescribed cost function. The following are some parameters that can be considered for inclusion in the optimization cost function:

- Distance traversed.
- Traveling on roads as far as possible even if the path becomes somewhat longer (road database).
- Avoiding paths with sudden changes in elevation. This factor is further influenced by the maximum slope the vehicle can negotiate (elevation or terrain database).
- Avoiding frequent changes in direction.
- Avoiding regions with forest cover or residential areas (land cover database).

In our experience, augmenting publicly available data with more detailed and accurate information about routes that are easily traversed is advantageous and will allow the vehicle to travel at higher speeds since we are more confident about the precise latitude and longitude of the roads [1, 3].

It should also be noted that, in production applications, these datasets are highly compressed and optimized for the needed tasks both to reduce storage and memory space and to reduce computational complexity, as the software may otherwise spend a considerable amount of time simply reading and parsing the data.

6.2.2   An Off-Road Grid-Based Path Planning Algorithm

The algorithm for path planning has to be implemented on a region within the overall map database. An A* search algorithm is used to perform path finding since it is very flexible and can be applied to the terrain map data, which is available in grid format. A* search is primarily a graph-based search method derived from Dijkstra's algorithm by employing a distance + cost heuristic function to determine the order in which nodes are searched. In this case, each tile on the grid can be viewed as a vertex and the line connecting two adjacent tiles can be viewed as an edge. Movement is possible only between vertices connected by edges. In this grid representation, each tile has eight adjacent tiles and vertices.

The A* algorithm maintains two sets, the OPEN list and the CLOSED list. The OPEN list keeps track of those vertices that need to be examined, while the CLOSED list keeps track of vertices that have already been examined. Initially, the OPEN list contains just the initial node and the CLOSED list is empty.

Each vertex n has a cost and a pointer associated with it. The pointer indicates the parent to vertex n, so that once a solution is found the entire path can be retrieved. In an A* search the cost associated with vertex n consists of two components:

$g(n)$: The cost of traveling from the initial vertex to n;

$h(n)$: The heuristic estimate of the cost of traveling from n to the goal vertex.

The total cost of vertex n is

$$f(n) : g(n) + h(n) \tag{6.1}$$

A* has a main loop that repeatedly obtains a vertex n with the lowest $f(n)$ value from the OPEN list. If n is the goal vertex, then by backtracking from n to its parent the solution is obtained. Otherwise, the vertex n is removed from the OPEN list and added it to the CLOSED list.

For each successor vertex n' of n, if n' is already in the OPEN or CLOSED list and its cost is less than or equal to the $f(n')$ estimate then the newly generated n' is disregarded. However, if the $f(n')$ estimate for n' already in the OPEN or CLOSED list is greater than the newly generated n' then the pointer of n' is updated to n and the new $f(n')$ cost estimate of n' is determined by:

1. Set $g(n')$ to $g(n)$ plus the cost of getting from n to n';
2. Set $h(n')$ to the heuristic estimate of getting from n' to the goal node;
3. Set $f(n')$ to $g(n') + h(n')$.

If vertex n' does not appear on either list then the same procedure is followed to set a pointer to the parent vertex n and to calculate the $f(n')$ estimate of the cost. Eventually n' is added to the OPEN list and returned to the beginning of the main loop.

The central idea behind A* search is that it favors vertices that are close to the starting point and vertices close to the goal. The A* algorithm balances the two costs $g(n)$ and $h(n)$ while moving towards the goal. The heuristic function gives the algorithm an estimate of the minimum cost from the current vertex to the goal. Hence, it is important to choose a heuristic function that can give the most accurate estimate. If the heuristic function is lower than (or equal to) the cost of moving from vertex n to the goal, then A* search guarantees the shortest path. However, if the heuristic estimate is greater than the actual cost, then the resultant path is not shortest, but the algorithm may execute faster. Some of the popular heuristic functions used in the case of grid maps are listed in the following subsections.

6.2.2.1 Manhattan Distance

This is a standard heuristic and its value is equal to the minimum cost of moving from one space to an adjacent space as described in Figure 6.5.

$$h(n) = |x - x_{goal}| + |y - y_{goal}| \qquad (6.2)$$

6.2.2.2 Diagonal Distance

This is used in grids that allow for diagonal movement as shown in Figure 6.6. The value of this heuristic assumes that the cost of moving along a straight line and along the diagonal is the same.

$$h(n) = \max\left(|x - x_{goal}|, |y - y_{goal}|\right) \qquad (6.3)$$

However, it can be modified to account for the difference in costs for moving along the diagonal.

$$\begin{aligned}
h\_diagonal(n) &= \min\left(|x - x_{goal}|, |y - y_{goal}|\right) \\
h\_straight(n) &= |x - x_{goal}| + |y - y_{goal}| \\
h(n) &= \sqrt{2} * h\_diagonal + (h\_straight - 2h\_diagonal)
\end{aligned} \qquad (6.4)$$

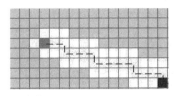

Figure 6.5 Illustration of Manhattan distance.

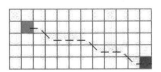

Figure 6.6 Illustration of diagonal distance.

6.2.2.3 Euclidean Distance

This heuristic is used when the robot can move in any direction and does not depend on grid directions. This value corresponds to the straight-line value of the path as shown in Figure 6.7 and the value is lower than that of the Manhattan or diagonal distance.

$$h(n) = \sqrt{\left(x - x_{goal}\right)^2 + \left(y - y_{goal}\right)^2} \tag{6.5}$$

Some implementation of A* algorithm employ the square of Euclidean distance. However, this may result in scaling problems as the value of the heuristic might become much larger than the value of $g(n)$.

In the case of path planning in uneven terrain, there are some regions where the terrain is completely inaccessible due to elevation, and some regions may be considered more favorable than others. In order to consider all possible terrain features while path planning, this aspect can be incorporated into the cost associated with the vertex.

6.2.2.4 Example Cost Function from the DARPA Grand Challenge

This section discusses the cost function used for an off-road autonomous vehicle deployed in the DARPA Grand Challenge. The problem can be simplified by considering only the following features for the cost function of a vertex:

1. Change in elevation – slope of terrain between the two vertices.
2. Greatest slope the vehicle can comfortably climb.
3. Distance between the two vertices.

Then the cost [4] associated with a vertex can be determined using

$$Cost(n, n') = f(d) * e^{s\lambda/\beta} \tag{6.6}$$

where the cost of traveling from vertex n to n' is a linear function of the distance and increases exponentially depending on the slope. The other constants appearing in (6.6) are:

s: slope, change in elevation from vertex n to n';

λ: slope weight—this constant prevents the exponent value from becoming very large;

d: distance between two points;

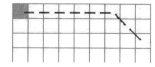

Figure 6.7 Illustration of Euclidean distance.

$f(d)$: Euclidean distance between n and n' in this case;

β: greatest slope the vehicle can comfortably climb.

In addition to the above terrain features, we can consider the presence of known good paths or roads. These features can be extracted from the appropriate map data layer. Since it is most favorable to travel on roads, a negative cost can be associated with roads in order to reduce the cost of a vertex if it lies on a road. This would lead to the algorithm automatically preferring roads over any other terrain type and irrespective of the slope.

Another possible cost function, which generates a smaller range of values, is given by the pseudo code

```
If road exists then
    Cost = Cost(n, n') - 2.00
Else if Elevation Change > 4.5 meters then
    Cost = arbitrary large integer
Else Cost = f(d)*|change in elevation|
```

Figure 6.8 shows a MATLAB rendering of the above cost function for a sample terrain map. Figure 6.9 illustrates a sample path planned over actual dessert terrain in southern California. A sample flowchart for the overall path planning algorithm is shown in Figure 6.10.

6.2.3 Other Off-Road Path Planning Approaches

In addition to the grid map strategy defined above, there are other possible algorithms available. These have been used in robotic systems for some time.

One approach is based on the extraction of a Voronoi diagram. A Voronoi decomposition of a set of points S in space is defined as the set of regions, each region containing one point in S, denoted s', and all other (empty) points that are closer to s' than any other point in S. A more direct and useful, at least for path planning applications, definition of a Voronoi diagram is a graph consisting of edges such that the edges are composed of points that are equidistance from objects, such as that shown in Figure 6.11. That is, if one constructs a Voronoi diagram from a list

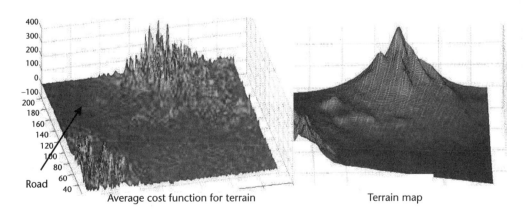

Figure 6.8 MATLAB simulation of modified cost function and elevation map for terrain.

Figure 6.9   Sample planned off-road route.

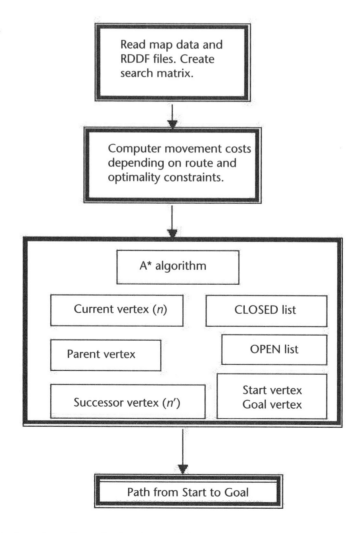

Figure 6.10   Flowchart of a path planning module.

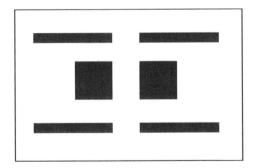
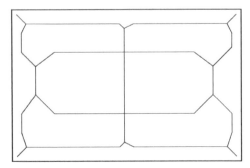

Figure 6.11   Sample obstacle map and the resulting Voronoi diagram.

of points or obstacles, a grid map, or any other representation of the data, then a vehicle or robot following an edge path is as far away from objects as possible. One must still plan a path through the resulting graph, using for example the A* algorithm described above.

Another approach to navigation and path planning involves treating all objects as if they have an electrical charge. The vehicle also has a charge of the same polarity as objects, and the goal is given a charge of opposite polarity. This scheme produces a vector field with the goal location at the minimum of the field. Various local or global optimization techniques can be used to find a path through the valleys of the field. However, potential field methods can be complicated. For example, many of the numerical optimization techniques may allow the solution to be trapped in a local minimum, thus never reaching the final goal.

6.2.4   An On-Road Path Planning Algorithm

A map of a city is essentially a network of many intersecting roads, and a map database serves as a digital description of the urban area in the real world. In particular, it specifies the network of road segments and unstructured free-travel zones accessible to the autonomous vehicle in a systematic fashion. The specific details of the description vary depending on the map database format used.

A representative example is the Route Network Definition File (RNDF) used during the DARPA Urban Challenge [5]. In this format, the route network consists of a number of road segments, each having one or more lanes. A lane is uniquely defined by a series of waypoints which are given by latitude and longitude coordinates. Intersections are defined by pairs of exit and entry points. Vehicles can exit from the exit point of one road segment and enter another road segment through the corresponding entry point. The RNDF also provides information such as the locations of stop signs and checkpoints, lane widths, and lane markers. In addition to road segments, the RNDF specifies free-travel zones to represent parking lots and obstacle fields. The zone area is restricted within a polygonal boundary defined by perimeter points. A zone may include one or more parking spots, each specified by a pair of waypoints.

There are a number of available map database formats, including the U.S. Census Tiger datasets, the Keyhole Markup Language (KML) from Google, the Geographic Markup Language (GDL) from the Open Geospatial Consortium,

Geographic Data Files (GDF), an international standard formalized by the European Committee for Standards (CEN), and the Standard Interchange Format (SIF) from Navteq.

6.2.4.1   Extraction of Graph Representation

The road network can be interpreted as a directed graph $G(N, E)$, where N corresponds to the set of nodes (vertices), and E is the set of edges. After appropriate translation from the route network information to a graph structure, existing shortest path algorithms such as the A* algorithm can be utilized to solve the route planning problem.

An intuitive graph model would represent all intersections by nodes and all streets by edges. However, the actual construction of the graph turns out to be much more complicated. One potential problem is that some intersections may have restrictions on turns, for example, prohibiting left turns. In constructing our graph model, we can consider exit and entry points, as well as any special points of interest noted in the map database, as the nodes of the graph. The edges between the nodes of the graph are defined with respect to available lanes, exit/entry pairs, lawful lane changes, and U-turns.

Following this logic, the graph representation of an intersection contains multiple nodes corresponding to the exit and entry points. The edges between them indicate admissible transitions from one road segment to another with proper left or right turns. This modeling scheme embraces all possible situations at intersections. Connections between different road segments can be defined appropriately by the edges in a consistent and straightforward way. Figures 6.12 and 6.13 illustrate a T-junction and a four-way intersection and their graph models. In Figure 6.12 only a right-turn exit/entry pair is defined, that is, vehicles on the horizontal road are not allowed to turn left onto the vertical road. If we modeled the T-junction with only one node, the no-left-turn policy could not be correctly conveyed by the graph. For free-driving zones, edges are assigned from entries of a zone to all nodes inside, and from both types of nodes to zone exits. Figure 6.14 exemplifies a graph model associated with a zone.

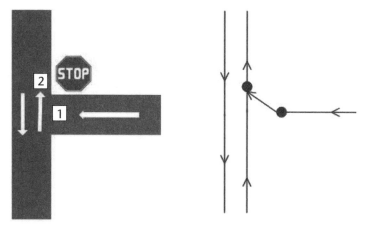

Figure 6.12 Graph model of a T-junction. Only one exit/entry pair (1, 2) is defined for the right turn. A left turn onto the left lane on the vertical road is not allowed.

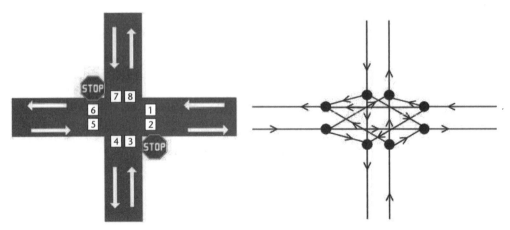

Figure 6.13   Graph model of a four-way intersection.

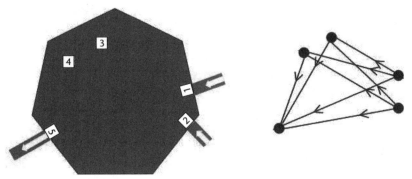

Figure 6.14   Graph model of a zone. Edges are established from zone entries to all nodes inside, and from these nodes to zone exits.

Our graph definition accurately describes the intricate urban road environment with consistency and flexibility, and therefore establishes a model for the route planner to utilize. Also, it enables the route planner to achieve a more accurate estimation of traveling time costs for the edges. What is more, the generated optimal routes can be represented in terms of series of waypoints, which reduces the effort expended interpreting and executing the routes in the control module.

6.2.4.2   Route Planning

One of the key challenges for an on-road autonomous vehicle is the optimal route planning for designated missions along with the capability of handling a dynamically changing road network through real-time replanning with updated road information [6]. For our purposes, we can consider a mission to be defined by a set of destinations in the graph representation of the road network. For simplicity we assume that the destination list is ordered.

The route planning module we describe is designed to calculate optimal routes for the vehicle for any given mission. The vehicle plans a route only from one destination to the next, with possibly a two-destination horizon. In certain situations, the route planner must generate routes originating from the vehicle's current position rather than a specific known node of the graph. In this case, the route planner has to locate the vehicle coordinates on the graph first, that is, to find the nearest approachable edge for the vehicle. Then it can plan a route from that edge. If the route planner is notified of a road blockage, it needs to update the graph structure before performing further searching tasks. Figure 6.15 illustrates the flow chart of the route planner.

6.2.4.3   Locating the Vehicle on the Graph Network

There are situations when the vehicle is not at a node of the graph and needs a plan from its current coordinates to the next checkpoint. For example, due to road blockage or other possibilities the vehicle may be unable to follow the original route at some point and need to resume its mission from that point. With the GPS-INS positioning system, the latitude and longitude coordinates and the vehicle orientation are available to the route planner. The planner should direct the vehicle to the

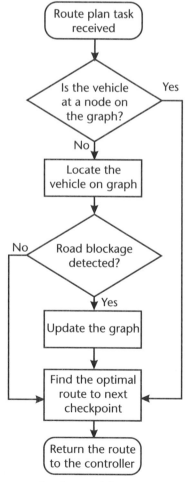

Figure 6.15   Flowchart of the route planning module.

nearest approachable edge with respect to its kinematical capability and then produce a route from that edge to continue with the mission.

To find the nearest reachable edge for the vehicle we need to search the graph with multiple criteria. We want to minimize the distance from the vehicle position to the edge as well as the deviation of the vehicle heading from the direction of the edge.

Consider an arbitrary edge E_{ij} on the directed graph. As shown in Figure 6.16, point P represents the position of the vehicle and vector $\overrightarrow{PN}$ indicates its heading. θ_1 is the angle between E_{ij} and $\overrightarrow{PN}$. The angle between E_{ij} and $\overrightarrow{Pj}$ is denoted by θ_2. We confine our search scope to edges with $|\theta_1| < 90°$ and $|\theta_2| < 90°$. To determine whether edge E_{ij} is a good match for our vehicle, we need to consider both how far away P is from E_{ij} and how well the vehicle orientation is aligned with the edge.

The distance from point P to edge E_{ij} is defined as

$$d\left(P, E_{ij}\right) = \min\left\{|\ \ |,\ \in\ _{ij}\right\} \tag{6.7}$$

We define a distance threshold D such that all edges E_{ij} with $d(P, E_{ij}) < D$ are considered nearby edges. An optimization function is then designed as:

$$f\left(P, E_{ij}\right) = \frac{\sqrt{D^2 - d\left(P, E_{ij}\right)^2}}{2D} + \frac{1}{2}\cos\theta_1 \tag{6.8}$$

This objective function leads to a weighted combination of distance and alignment optimization. The two parts in the function are convex with respect to $d(P, E_{ij})$ and θ_1, respectively. $f(P, E_{ij})$ takes a value between $[0,1]$ and achieves its maximum when $d(P, E_{ij}) = 0$ and $\theta_1 = 0$.

We also need to consider the minimum turning radius of the vehicle, determined by [7]

$$\rho_{\min} = \frac{L}{\tan\phi_{\max}} \tag{6.9}$$

where L is the distance between the front and rear axles of the vehicle and $\phi_{\max}$ is the maximum steering angle. If the vehicle stands too close to the end node of an

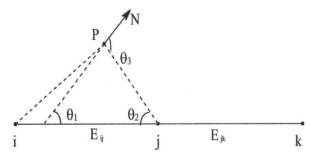

Figure 6.16 Edges E_{ij} and E_{jk} on the graph and vehicle position P with heading PN.

edge it cannot approach that node without backing up. However, reversing on the road is not allowed. We can determine whether the node is reachable for the vehicle by the following proposition.

Assume that a vehicle is standing at point P with an orientation $\overrightarrow{PN}$ and there is a nearby edge E_{ij}. The vehicle has a minimum turning radius of ρ_{min}. Fix point O_1 such that PO_1 is normal to $\overrightarrow{PN}$ and $|PO_1| = \rho_{min}$ and also fix point O_2 such that jO_2 is normal to E_{ij} and $|jO_2| = \rho_{min}$ as shown in Figure 6.17. If the angle between $\overrightarrow{PN}$ and edge E_{ij} is acute and $|O_1O_2| \geq 2\rho_{min}$, then the vehicle can arrive at node j with a proper nonreversal maneuver.

We use this fact to decide whether edge E_{ij} of the search results is approachable. If not, we check the successors E_{jk} of that edge and choose the one with maximum $f(P, E_{ij})$ as the final result. In short, the searching algorithm works as follows:

- Step 1: In graph $G(N, E)$, find

$$E^1 = \underset{E_{ij} \in E, |\theta_a| < 90°, |\theta_2| < 90°}{\text{«««}} d\left(P\ E_{ij}\right) \tag{6.10}$$

- Step 2: In the set E^1 of edges found in step 1, calculate

$$E^2 = \underset{E_{ij} \in E^1}{\text{«««}} d\left(P\ E_{ij}\right) \tag{6.11}$$

- Step 3: Check whether the edge E^2 is reachable by the vehicle. If not choose an edge E^3 with the largest $f(P, E_{jk})$ from its successors. The edge found is the best match for the vehicle.

6.2.4.4 Updating the Graph

A road may become blocked due to severe traffic jams, road construction, traffic accidents, or vehicle failures. An autonomous vehicle should be able to react to such scenarios and replan its route. When the route planner receives notification of a road blockage it needs to update the nodes and edges in the graph. First, the cost of the related edges can be increased to a predefined high value C_{blk}, which is an

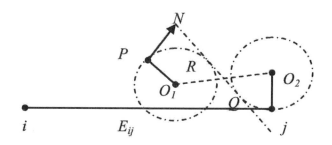

Figure 6.17 Construction of a feasible trajectory from P to node j.

estimate of the cost of sticking to the original plan and thus remaining stationary while waiting for the road to clear. This high cost leaves the original plan as the last yet still possible option.

Second, new nodes and edges might need to be added to the graph to allow U-turns on both sides of the blocked road segment. It is also possible that the destination might be unreachable since it is on the road segment past the road blockage. As an alternative to simply waiting for the blockage to clear and continuing with the original plan, the vehicle could navigate in order to reach the far road segment beyond the blockage and then reach the destination by a U-turn maneuver as shown in Figure 6.18. To incorporate a U-turn edge into the defined graph on the destination side of the barrier, the planner can identify the lane of the destination and find a lane next to it with opposite direction. On this later lane the planner searches for a waypoint located closest to the destination, or defines it as a node if a waypoint does not exist and establishes an edge between this node and the checkpoint corresponding to the U-turn.

It is reasonable to assume that the road blockages will be cleared after a certain length of time T_{blk} and the graph can update itself again for later planning. The planner keeps records of all blocked segments with time stamps and releases a record and undoes the corresponding modifications after T_{blk} seconds have passed since the establishment of the record.

6.2.4.5 Searching for the Optimal Route Between Two Nodes

We can design the route planning module to support both shortest-distance and shortest-time route searching. Accordingly, two schemes of cost definition for the edges are determined based on traveling distance and time, respectively. A distance-based cost is the physical distance across the two ends of the edges. A time-based cost is an estimate of time for the vehicle to drive through the edges. Because the vehicle does not necessarily maintain a constant speed along the way, the minimum-distance route and the minimum-time route can be different.

For a time-based cost system the accuracy of estimates in driving time is crucial. It is not possible to derive exact driving time for each edge due to uncertainties about the road condition and the traffic. However we can still acquire reasonable estimates by taking into account the expected speeds of the vehicle over the edges as well as stop signs and intersections and the possible U-turn and obstacle avoidance maneuvers associated with them.

We can calculate the time cost of an edge based on the vehicle speed and the expected delay with stop signs or U-turn maneuvers associated with the edge using

Figure 6.18 Definition of new edges for U-turns at the blockage.

$$t\left(E_{ij}\right) = \frac{L\left(E_{ij}\right)}{v\left(E_{ij}\right)} + t_d \tag{6.12}$$

where $L(E_{ij})$ is the physical length of E_{ij} and $v(E_{ij})$ is the expected average driving speed the vehicle uses for the edge. This can depend on a speed limit extracted from the map database, some indication of the type of road, the length of the segment, and the presence of intersections. The term t_d is the expected time delay due to stop signs or U-turn maneuvers. It can be constant for simplicity or a function of the vehicle speed to achieve higher accuracy.

Figure 6.19 illustrates a minimum-time route on an example map. The vehicle starts from node S and is to arrive at the destination node T within the shortest time. Therefore the path planner would choose the highlighted route which runs through nodes A, B, and C over the other candidate route with nodes D, E, F, G, H, J, and C, which is shorter in distance but involves two more intersections.

A* searching [8] is known as an efficient and effective algorithm for the shortest path searching problem with a single source and single destination. The key element of A* is the "heuristic estimate" $h(n)$ for estimating the so called cost-to-go, which is the cost from node n to the destination. To ensure that the optimal solution always be found, $h(n)$ has to be admissible and consistent [9]. When aiming at a minimum-distance route, our $h(n)$ is the straight-line distance from node n to the destination. This distance heuristic $h(n)$ fulfills the requirements of [9]. For optimality in time, we define our $h(n)$ as the straight-line distance from node n to the destination divided by the upper bound of maximum speed limits over the network. As a scaling of the distance heuristic, this preserves the properties of admissibility and consistency.

The implementation of an A* algorithm to determine a minimal cost route from a start node s to an end node t follows the logic below. The module maintains two sets of nodes, an open set P and a closed set Q. To estimate the lowest cost of routes passing through node n, the function $f(n) = g(n) + h(n)$ is defined for all nodes n in the open set P. Here $g(n)$ is the lowest cost of the routes from s to n with the routes only consisting of nodes in set Q and $h(n)$ is an estimate of the cost-to-go. Initialize set P to contain only the start node s and Q to be empty. At each step, move the node n with the smallest $f(n)$ value from P to Q, add the successors of n to set P if they are not already present, and update $f(n)$ for all these successors in

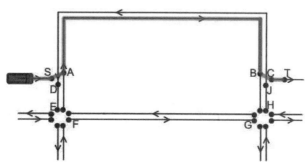

Figure 6.19 An example of minimum-time route.

P. Repeat the operation until *t* is included in *Q* to obtain the optimal route. If *P* becomes empty before that, then no route from *s* to *t* exists.

The dimension of the graph model grows rapidly with the scale of the road network in terms of the number of roads and intersections. Therefore, it is very important that the route planner is economical in computational resources and data structures.

When the optimal route is found, the route planner should compile the plan and send it to the vehicle's behavior and high-level control algorithms. Note that it is possible to identify the behavior required to execute each section of the chosen path. These may be the metastates in a hierarchical behavior control system (i.e.,

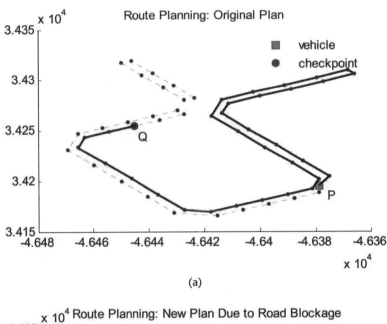

(a)

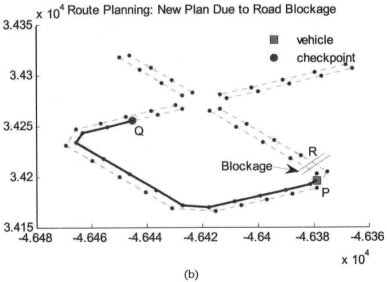

(b)

Figure 6.20   (a, b) Sample routes in a simple road network. The dashed lines imply the center lines of the available lanes, with waypoints marked by dots.

lane following, intersection processing, and so forth) that define the required maneuvers and therefore the corresponding vehicle control strategies [10].

6.2.4.6 Example

The example shown in Figure 6.20 demonstrates the capability of solving planning problems in a dynamically varying route network with successful real-time replanning. The road network consists of a traffic loop and two extended stubs with U-turns permitted at the ends of the stubs. The vehicle at point P was heading for the destination checkpoint Q when it found the blocking barrier in its way at point R. An alternative route to the destination was calculated, which happened to be much shorter thanks to the U-turn allowed at the blockage. Both the original route and the new plan are shown in the figure.

References

[1]   Toth, C., et al., "Mapping Support for the OSU DARPA Grand Challenge Vehicle," *Proc. 2006 IEEE Intelligent Transportation Systems Conference*, Toronto, Canada, September 17–20, 2006, pp. 1580–1585.

[2]   LaValle, S. M., and J. J. Kuffner, "Randomized Kinodynamic Planning," *International Journal of Robotics Research*, Vol. 20, No. 5, May 2001, pp. 378–400.

[3]   Dilip, V., "Path Planning for Terramax: the Grand Challenge," M.S. Thesis, The Ohio State University, 2004.

[4]   Marti, J., and C. Bunn, "Automated Path Planning for Simulation," *Proc. of the Conference on AI, Simulation and Planning (AIS94)*, 1994.

[5]   Defense Advanced Research Projects Agency, "Urban Challenge," http://www.darpa.mil/grandchallenge.

[6]   Fu, L., A. Yazici, and Ü. Özguner, "Route Planning for the OSU-ACT Autonomous Vehicle in DARPA Urban Challenge," *Proc. 2008 IEEE Intelligent Vehicles Symposium*, Eindhoven, the Netherlands, June 4–6, 2008, pp. 781–786.

[7]   Dubins, L. E., "On Curves of Minimal Length with a Constraint on Average Curvature, and with Prescribed Initial and Terminal Positions and Tangents," *American Journal Mathematics*, Vol. 79, July 1957, pp. 497–516.

[8]   Hart, P. E., and N. J. Nilson, "A Formal Basis of the Heuristic Determination of Minimum Cost Paths," *IEEE Transactions on Systems Science and Cybernetics*, Vol. 4, No. 2, July 1968, pp. 100–107.

[9]   Russell, S. J., and P. Norvig, *Artificial Intelligence: A Modern Approach*, Upper Saddle River, NJ: Prentice-Hall, 1995, pp. 97–104.

[10]   Kurt, A., and Ü. Özguner, "Hybrid State System Development for Autonomous Vehicle Control in Urban Scenarios," *Proc. of the IFAC World Congress*, Seoul, Korea, July 2008.

CHAPTER 7

Vehicle-to-Vehicle and Vehicle-to-Infrastructure Communication

7.1 Introduction

Advances in wireless technologies are providing rapid increases in the availability of mobile data to passengers and drivers, intelligent vehicles and transportation applications. Cell phones and smart phone applications are ubiquitous and thus an obvious example, but do not provide solutions for all applications. It is necessary to select a communications technology that matches the requirements of a specific application.

There are many ways to classify the wireless applications and interactions that occur among vehicles or between vehicles and the infrastructure. One classification would be based on their purpose and information contents, for example safety systems, vehicle control systems, roadway operation systems, driver information and convenience systems, and recreation, pleasure, or luxury applications. Another classification could be based on technical requirements such as:

- *Digital bandwidth:* The rate at which data can be transferred;
- *Latency:* The delays inherent in moving data from source to destination;
- *Reliability:* The likelihood that a message will be received at its destination;
- *Security and authentication:* The ability to verify the identity of the source of a message and to guarantee that its contents has not been tampered with or read by an unauthorized receiver;
- *Network configuration:* For example, broadcast, broadcast with forwarding, or point-to-point routed communication topologies.

Finally, one can classify based on the intended communication participants, such as other nearby vehicles, for example, all those vehicles interacting at a specific intersection; vehicles further away, for example, all vehicles along a stretch of freeway; vehicle and roadway or transportation infrastructure components; or vehicles and remote data source, for example, the World Wide Web.

A summary of communication technologies is given in Table 7.1, which illustrates the ranges of applications, data rates, and communication distances that various technologies may provide.

Generally speaking, the most unique characteristics of a vehicular communication network are the mobility of each agent and the potential reliability requirements of some applications. Agents may be moving at relative speeds exceeding 250 km/hour (approximately 70 m/s). This provides a series of technical challenges in the radio and medium access layers of the communication devices. It implies that the network, the collection of vehicles that are expected to communicate, can change both in size and composition at a very rapid pace. It also implies that some of the data is extremely time critical so that routing, acknowledgment, or retransmission protocols may not be feasible.

These unique characteristics have led to specific terminology. A vehicular ad hoc network (VANET) is considered to consist of vehicles and potentially road side units (RSU) equipped with wireless interfaces providing communication among cars [vehicle-to-vehicle communications (V2V)] and the infrastructure [vehicle-to-infrastructure communications (V2I)]. A VANET is a particular subset of mobile ad hoc networks (MANET), which are characterized by high speeds, persistent mobility, and mostly short-lived connections (e.g., a vehicle passing another vehicle stopped on the emergency lane). Short range wireless interfaces collaborate to form a temporary distributed network enabling communications with other vehicles and RSUs located within line of sight and even outside the direct unit-to-unit communication range by utilizing multiple hops through vehicular gateways. As an example, consider information propagation about an accident scenario as illustrated in Figure 7.1.

As the technology has advanced, there have been a number of large government-sponsored research programs exploring various aspects of vehicle-to-vehicle and vehicle-to-infrastructure communications.

Table 7.1 Summary of Communication Technologies and Applications

Application	Standards	Data Rates	Range
ECU, vehicle control bus, sensors	CAN, LIN, FlexRay	10 Kbps –10 Mbps	In vehicle only (direct connection)
Multimedia	MOST, IEEE1394b	20–1,000 Mbit/sev	In vehicle only (direct connection)
Cooperative ACC, intersection, hazard warning, platoon, merging (low latency)	DSRC 802.11p/WAVE	500 Kbps–27 Mbps	500 meters, vehicle-to-vehicle, low latency
Traffic information, roadside information	DSRC 802.11p/WAVE	500 Kbps–27 Mbps	500 meters, vehicle-to-vehicle, high latency, multihop
Tolling, intersection signal information, localized traffic information	DSRC 802.11p/WAVE	500 Kbps–27 Mbps	500 meters, vehicle-to-infrastructure
Traffic information, weather, GPS correction services	FM subcarrier, digital and satellite radio	120 bps–6 Mbps	50 km, vehicle-to-infrastructure unidirectional
Internet access, streaming media	Digital cellular, CDPD, GSM, and so forth	9,600 bps–6 Mbps	Effectively unlimited, vehicle-to-infrastructure

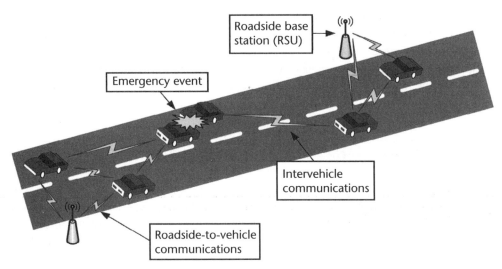

Figure 7.1 Aspects of a VANET.

In the United States, the National Vehicle Infrastructure Integration Initiative (VII), sometimes referred to as Intellidrive, is a cooperative effort between the U.S. Department of Transportation [1], various state departments of transportation, and the automobile industry to develop and test an information infrastructure that uses the most advanced communications technologies to exchange real-time information between the roadside and vehicles to improve safety and mobility [2]. Dedicated short range communication (DSRC) equipment operating in the 5.9-GHz frequency range is placed on the roadways and within the vehicle. Specific applications are being developed to test a broad variety of potential safety and mobility uses of the VII system including:

- Warning drivers of unsafe conditions or imminent collisions;
- Warning drivers if they are about to run off the road or are approaching a curve at too high a speed;
- Informing transportation system operators of real-time congestion, weather conditions, and incidents;
- Providing operators with information on corridor capacity for real-time management, planning, and provision of corridor-wide advisories to drivers.

The VII consortium is also considering issues related to business models, legal issues, security and technical feasibility, and acceptance among the various stakeholder parties.

In the European Union, a number of research activities have been coordinated under the eSafety Program [3, 4], including the Car2Car Consortium [5], COMeSafety [6], and the Cooperative Vehicle-Infrastructure Systems [7] program, among many others [8, 9]. In Japan, various programs have used V2V communication of some form, including the Advanced Cruise Assist Highway Safety Research Association (AHSRA), the Active Safety Vehicle (ASV3) programs, the Advanced Vehicle Control Systems (AVCS), and ITS Energy activities [10–12].

In this chapter we will summarize vehicle-to-vehicle and vehicle-to-infrastructure communications applications, approaches and technologies, and potential applications for autonomous vehicles. Perhaps the most general classification of the application of wireless communications in vehicles is to distinguish between vehicle-to-vehicle and vehicle-to-infrastructure applications, and we will discuss both in succeeding sections.

7.2 Vehicle-to-Vehicle Communication (V2V)

Direct communication between vehicles allows information exchange without requiring any fixed infrastructure or base stations. The location and velocity of vehicles is constantly changing, and the RF communication range is of fairly short distance; therefore, the set of vehicles that can directly communicate will constantly change over a short period of time. This dictates that the physical layer and the network must be capable of operating in an ad hoc, decentralized manner, although coordination and synchronization through GPS time signals are possible. Any two nodes must be able to communicate securely whenever they are within communication range.

In a V2V network we can distinguish two modes of communication, usually designated as:

- *Single hop:* Two vehicles are close enough to communicate directly with each other (either broadcast or point to point) with low latency.
- *Multihop:* Vehicles that can not directly communicate may forward messages through intermediate nodes.

Multihop communication has been the subject of much research [13], but no standard has emerged, and in fact the technical difficulties of establishing routing and acknowledgment protocols along with potentially high latency may limit its use to very specific applications such as medium range emergency notification or other sparse broadcast communication applications.

Many early experiments in V2V communication were carried out with standard wireless LAN technologies, for example IEEE 802.11b, operating in the 2.4-GHz ISM band, and some success was achieved at ranges of up to several hundred meters. But the technical difficulties inherent in vehicle and traffic situations, including the high relative velocities (Doppler effects), a safety critical low latency requirement, operation in an urban environment (multipath), and spectrum competition from other users in unlicensed frequency bands renders this an unrealistic solution for commercial deployment. The IEEE 802.11p/WAVE standards have recently emerged as the current consensus for the implementation of V2V and local V2I communications. They will be described later in this chapter.

One very important issue in the use of DSRC technology is the level of market penetration. The V2V technology will not provide a benefit to those who have it until a sufficient number of vehicles have been equipped with the sensors, communication technology, processing, and human interfaces required for the applications. The V2I technology faces a similar hurdle, but in addition it requires government agencies or other providers to commit the resources to deploy a sufficient number of roadside and infrastructure units.

There are a number of example applications that are made possible using V2V communications alone. The most obvious are the active safety systems, which aim to prevent or mitigate possible hazards by generating driver warnings or by direct intervention in the behavior of the vehicle. In these applications, communication between vehicles or roadside units will increase perception limits by propagating the information through mobile or stationary nodes. Wireless information may decrease the reaction delays of the driver by allowing for earlier preparation and increased vigilance to an oncoming traffic scene.

From the application viewpoint, safety-related examples include:

- Stopped vehicle or obstacle avoidance on highways (see Figure 7.2);
- Merging assistance from a slow incoming lane into a crowded road (see Figure 7.3);
- Intersection safety and collision warning systems (with or without traffic lights) (see Figure 7.4).

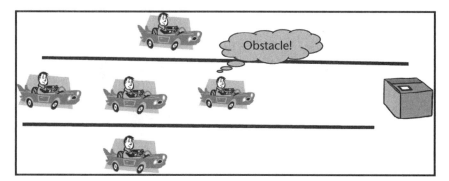

Figure 7.2   Obstacle avoidance in a multilane situation.

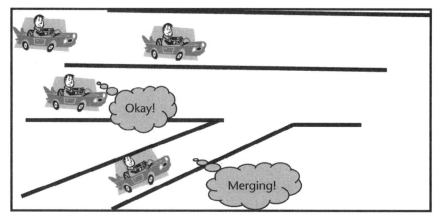

Figure 7.3   Merging into lane and gap search/creation.

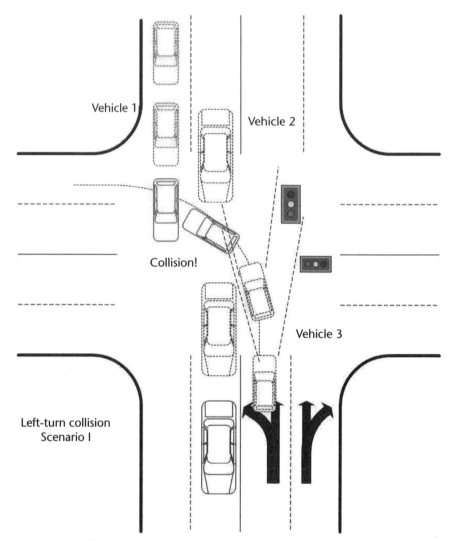

Figure 7.4 Potential intersection collisions: the left-turn case.

For example, in a vehicle-based intersection collision warning system, each vehicle approaching the intersection generates a message. The message includes the vehicle's location, direction of travel, and the intersection location. This information is based on the GPS position and the available road map database. Various strategies to predict driver intention are also being studied, and if available that too is broadcast. Any other vehicle approaching the intersection can calculate the possibility of a collision and will generate a warning for its driver if needed. This can also be expanded to include traffic signal state, in which case the application is utilizing both V2V and V2I communications [14].

There are, of course, many other possible applications. For example:

- Notification of the presence or approach of an emergency vehicle;
- Notification of construction activity;

- Local multivehicle sensor fusion to provide better global maps and situation awareness;

- Energy-saving applications, for example, platooning and energy management and control for HEVs in stop-and-go traffic.

7.3   Vehicle-to-Infrastructure Communication (V2I)

The term vehicle-to-infrastructure communication (V2I) is used for any kind of communication from the vehicle to a fixed infrastructure or vice versa. This communication can be unidirectional or bidirectional. Broadcast systems support unidirectional transfer of information from a broadcast station to the vehicle. In contrast, in systems allowing bidirectional communication the vehicle communicates point-to-point with a base station. In this case, the base station may be responsible for coordinating the communication (e.g., physical layer synchronization and medium access) in which case the base station can also provide access control and congestion management.

V2I can be provided through dedicated equipment, for example DSRC, which offers short range communications but a potentially high data rate, or through existing systems such as cellular mobile phone providers.

Perhaps the best known and most widely deployed V2I application is electronic toll collection (ETC), which provides an automatic deduction of toll and road usage charges from a prepaid account tied to the vehicle's wireless radio or transponder. For commercial vehicles this can be expanded to include cargo and licensing information and enforcement applications.

Active safety system applications include intersection safety systems with traffic signals that transmit their current state and cycle time information. A current major research initiative is the U.S. Cooperative Intersection Collision Avoidance System [14].

There are also numerous possible applications for traffic and traveler information systems, for example providing road network and map updates and traffic congestion information to an in-car navigation system or directly to the driver. Consider an "alternate route system," in which a congestion display could inform the driver of estimated delays in traversing routes. This type of display would be updated with wireless data provided from a central source, which collects all traffic information from all city arteries and would have local knowledge of the intended destination and primary route of the driver. The system in the car would then calculate the time to destination along several alternate routes. If, due to congestion on the current route, the system discovers routes that will require less time to traverse, it will provide this information to the driver. Targeted advertising is also sometimes suggested as an application.

7.4 Communication Technologies

A wide range of technologies have been considered, or in some cases deployed, for V2V and V2I communications. This section will describe some of those that have been actively implemented.

7.4.1 Unidirectional Communication Through Broadcast Radio

There are existing standards for digital communication using subcarriers of standard AM and FM radio broadcast stations. Applications include channel and programming information, commercial paging systems, weather, news, and traffic information, stock market quotes, and GPS differential correction services. The data rate is quite low (on the order of 1 Kbps) and the services often require paid subscriptions. Many of these applications are declining in popularity due to the availability of other, faster technologies. Satellite radio offers a similar unidirectional capability at much higher data rates, for example, Sirius Traffic, a subscription service for real-time traffic data.

7.4.2 Cellular/Broadband

As we have previously mentioned, cellular telephone and broadband data services have become ubiquitous. Pricing, at least for individual users, is still rather high, but vehicle OEM and other providers have negotiated pricing arrangements for particular services. Perhaps the best known is the GM OnStar service, which provides driver information including directions, stolen vehicle location, crash detection and emergency services notification, and other services. BMW Assist is a similar service. To date, these services have been implemented by specific vehicle manufacturers and are not available outside their vehicle brands.

7.4.3 Information Showers

The concept of installing dedicated, short range unidirectional communication systems at regular intervals along the roadway has been implemented in Japan under the Vehicle Information and Communications System (VICS) [15], a program of the Japanese Government. Real time traffic and other information is provided through a number of short-range communication technologies, including:

- *Infrared beacons:* 1 Mbps, 3.5-meter range;
- *2.5-GHz radio beacon:* 64 Kbps, 70-meter range;
- *FM multiplex:* 16 Kbps, 10–50-km range.

7.4.4 Narrowband Licensed 220 MHz

In 1992 the U.S. Federal Communication Commission temporarily allocated five narrowband frequency pairs in the 220–222-MHz band to the Federal Highway Administration for intelligent transportation system experimental and testbed applications. These allocations are capable of supporting low-rate digital communications

(4–30 Kbps). Several test deployments were made, including a traffic management application in Louisville, Kentucky. Research involving high spectral efficiency modulation schemes and software-defined radio development was carried out at Purdue University [16] and at The Ohio State University [17]. In December 2007, the channels were still allocated to FHWA but the only known operational system consisted of a Delaware DOT traffic monitoring application that was phased out in 2009 [18]. FHWA has since dropped support for the licenses. While the demonstrations and deployments were successful, and there were RF propagation advantages available in this lower frequency band, the extremely narrow channel bandwidth has generally proved impractical for large-scale deployment.

Recently, there have been requests to reallocate certain channels within the 700-MHz band for use by both public safety and intelligent transportation system applications to provide licensed high-speed broadband capability using a cellular-like network.

Figure 7.5 shows a demonstration application developed at OSU in 2003 for a long-distance information and warning system based on a 220-MHz software-defined radio (SDR).

7.4.5   Dedicated Short-Range Communication (DSRC)

DSRC systems are short- to medium-range communications systems intended to cover communication ranges of 10–1,000 meters. The term DSRC has come to refer to automotive or mobile applications. A number of technologies have been identified, but the current object of worldwide standardization activities are variants of the 802.11p/WAVE standard operating in the 5.9-GHz range. The United States has currently allocated 75 MHz of spectrum for DSRC applications, and the EU has allocated 35 MHz of overlapping spectrum. These standards will be described more fully in Section 7.5.

Other DSRC technologies include a 902–928-MHz band standard (ASTM E2158-01) that has primarily been used in electronic toll collection and commercial vehicle operation applications. It is incompatible with the 5.9-GHz DSRC standards.

A Japanese industrial consortium, including OKI Electronics Ltd., developed the "Dedicated Omnipurpose Intervehicle Communications Linkage Protocol for Highway Automation" (DOLPHIN) system operating in the 5.8-GHz band [19–21], which provided broadcast, point-to-point, and broadcast with packet forwarding capabilities.

Bluetooth, various UWB, WiMAX, and even Zigbee (IEEE 802.15.4) could also be considered for DSRC applications.

7.5   802.11p/WAVE DSRC Architecture and U.S./EU Standards

The United States, since 1999, and the European Union, since 2008, have allocated spectrum in the 5.9-GHz band for wireless DSRC systems for vehicular safety and intelligent transportation system applications. The standards for DSRC in the United States are IEEE 802.11p, finalized in 2010, and WAVE (IEEE 1609), of which 1609.3 and 1609.4 have been approved and 1609.1 and 1609.2 remained

(a)

(b)

Figure 7.5   (a, b) The OSU demonstration of V2V networking in 2003.

in draft or trial use status at the time of this writing. The 802.11p standard is derived from the 802.11a wireless LAN standard and thus provides a fairly easy and inexpensive route to produce communication electronics and hardware. These standards provide network and application support, as well as sufficient range and data rates for both V2V and V2I communication applications.

In addition to wireless communication standards, the development of standardized message sets and data definitions in the United States has occurred in the SAE J2735 standards process.

Wireless access in vehicular environments (WAVE) is comprised of multiple IEEE standards, including:

- IEEE 802.11p, which defines the physical and medium access layer;
- IEEE 1609.1, which defines resource, system, and message management as well as one possible interface to applications;
- IEEE 1609.2, which defines security and encryption services;
- IEEE 1609.3, which defines the network and transport layers and the new WAVE short message (WSM);
- IEEE 1609.4, which defines the channelization, which is unique to the WAVE stack, and message prioritization occurring in the MAC layer.

The relationship of these standards in WAVE to the network stack is shown in Figure 7.6.

7.5.1 802.11P Physical Layer

802.11p is an extension of the 802.11 standard and most closely resembles 802.11a. 802.11p uses an orthogonal frequency-division multiplexing (OFDM) modulation scheme in the 5.85–5.925-GHz range on seven different channels. The use of channels will be discussed in further detail in Section 7.5.2. 802.11p contains a similar physical layer convergence procedure (PLCP) to 802.11a/b/g. 802.11p does not require RTS/CTS handshaking, but does require acknowledgments for unicast packets and support for packet fragmentation. 802.11p uses OFDM with 48 subcarriers and 4 pilot carriers with a nominal bandwidth of 10 MHz, and uses multiple modulation schemes with OFDM to support bit rates from 3 Mbps to 27 Mbps. It also supports the use of randomized local MAC addresses.

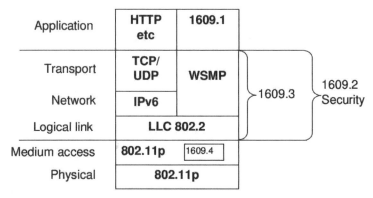

WSMP – WAVE Short Message protocol

Figure 7.6 Relationship of WAVE stack and standards.

The first task of the OFDM PLCP sublayer is to create a frame that can be easily demodulated at the receiver. This frame is called the PLCP protocol data unit (PPDU), a breakdown of which is shown in Figure 7.7.

A PPDU is created by building a preamble of fixed length (for 802.11p, this is 32 μs), which is used for synchronization at the receiver. Following the preamble is a PLCP header using binary phase shift keying (BPSK) and a fixed coding of $R = 1/2$. This header provides the receiver with coding rate, modulation type, and length of the rest of the message. The receiver can use these values to begin demodulating the message. The next field is the service field; it contains the scrambler seed that was used to "randomize" the message. This is important for descrambling the message and undoing the circular convolution.

7.5.2 1609.4 Channelization Overview

Above the 802.11p physical layer in WAVE is the 1609.4 channelization specification. This specification defines the use of a control channel (CCH) and six service channels (SCH). A WAVE device will be required to tune to the CCH channel at 100-ms intervals aligned with UTM second (for example, derived from a GPS receiver) or derived from the switching of another nearby transceiver, and monitor the CCH for messages for a set interval, after which the device is allowed to switch to a service channel. This channel switching scheme is shown in Figure 7.8. Another option is to use multiple transceivers so that one can be continuously allocated to the CCH and the other can be used on the SCH.

The CCH only allows a new protocol called WAVE Short Message Protocol (WSMP), an efficient alternative to IPv6 packets. These messages are approximately

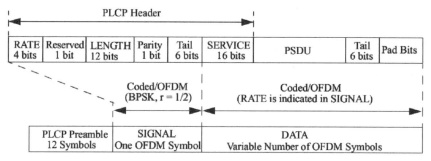

Figure 7.7 PLCP protocol data unit.

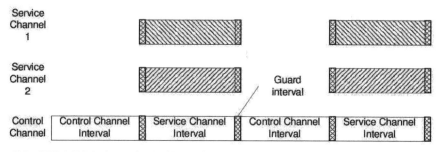

Figure 7.8 IEEE 1609.4 channel switching scheme.

400 to 500 bytes and have a new EtherType (0×02) to allow the WAVE network stack to handle them differently than the normal 802.2 frames. Service channels can use either IPv6 protocols or WSMP. Both the CCH and SCHs use a priority queuing scheme based on 802.11e EDCA.

Figure 7.9 shows the priority queues used on the channels. Each queue represents the access category index or ACI. Table 7.2 shows some of the parameters used for the CCH prioritization (the SCH use similar values). These values are used for determining which message will be sent next. This is useful if there are many queued regular priority messages to be sent and a high priority emergency message is created. The arbitration interframe space number (AIFSN) parameters indicate the minimum number of short interframe space (SIFS) windows a packet of a given priority will wait. This number will be added to a random number between 1 and the CW range when a packet reaches the head of the queue. For example, a CCH packet with a best effort priority could get assigned a wait time of 6 through 13.

The channel dwell intervals also require the 1609.4 section to query the physical layer (802.11p) regarding how much time will be required to send a selected message on the current channel. If the time required exceeds the remaining time for the current channel interval and thus would interfere with the guard interval, the

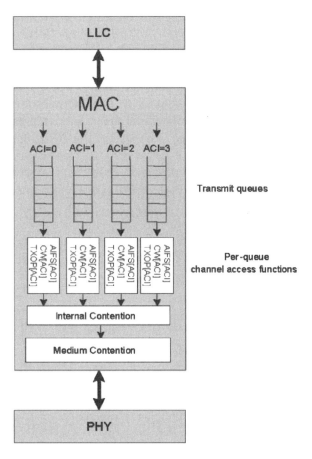

Figure 7.9   Prioritized access on control and service channels.

Table 7.2 EDCA Parameters for CCH Messages

ACI	AC	AIFSN	CW Range
1	Background	9	15
2	Best effort	6	7
3	Video	3	3
4	Voice	2	3

messages will be buffered in the MAC layer and transmitted later upon return to the correct channel. The default channel dwell interval is 50 ms.

7.5.3 1609.3 Network Management

The 1609.3 standard details how a WAVE system will respond to service advertisements, how priorities are assigned to each message, and how a WAVE device should offer services of its own. The 1609.3 is the network management layer that decides to which SCH to tune when the control channel interval is complete. It does this by interpreting and generating special WSMs called service announcements. A WAVE service information announcement (WSIE) is sent by a WAVE device that has a service available for use by another WAVE device. The WSIE is made up of the service type and the SCH number on which it will be available and an optional WAVE routing advertisement.

All applications that wish to provide or to subscribe to a service must register themselves with the 1609.3 WAVE management entity (WME). Each application will have an assigned priority and the higher priority has precedence when the 1609.3 layer is queried for the next SCH channel selection. During registration the application also provides information about service persistence and the number of message repeats. If a service is persistent, the WSIE message created for it will be sent whenever the control interval is entered on the CCH at the beginning of a UTC second. The number of repeats defines the number of times the same WSIE message must be sent during the control channel interval. Persistence and repeats are illustrated in Figure 7.10.

When an application wishes to subscribe to a service, it also registers itself with the WME. After that, whenever a WSIE is received, its service information is

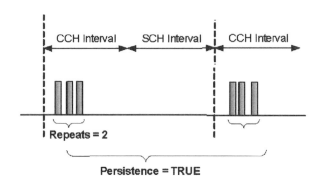

Figure 7.10 Repeats and persistence in WSA.

tested against the list of registered applications. If an application matches and has a higher priority then the currently active service, the SCH specified in the WSIE is set as the next active service channel.

Two types of WAVE devices are supported, the roadside unit (RSU) and the on-board unit (OBU). The RSU would typically be a service provider deployed on the roadside that would provide general information, for example, map data like the geometric intersection data (GID) message or local speed limits. If it is connected to a traffic light it could also provide signal phase and timing (SPaT) messages. Both message types are defined in [22]. The OBU is responsible for receiving data from RSUs and generating and receiving safety messages that relay information about the vehicle's current state.

7.5.4 EU Programs and Standards Activity

The European communication standards in the 5.9-GHz band are based on the published IEEE 802.11p standards. The ETSI standard ES 202 663, produced by the ETSI Technical Committee for Intelligent Transport System (ITS), describes several frequency ranges and the medium access control. A frequency band ranging from 5,895 GHz to 5,905 GHz is allocated as a control channel with a 10-MHz channel spacing for ITS safety-related applications. This functionality is denoted as the "G5CC ITS-G5 control channel." Four fixed physical channels are identified as the service channels and this functionality is denoted as G5SC. Each individual service channel has a 10-MHz spacing, and service channels in the frequency band ranging from 5,855 GHz to 5,875 GHz are dedicated to ITS road safety applications, whereas those in the frequency band ranging from 5,875 GHz to 5,895 GHz are dedicated to ITS roadsafety applications. An additional frequency allocation ranging from 5,470 GHz to 5,725 GHz is dedicated to ITS applications for possible use with radio local area networks (RLAN) and broadband radio access networks (BRAN).

The ETSI Technical Committee ITS standard EN 302 571 also defines the standardization of radiocommunications equipment operating in the frequency band ranging from 5,855 MHz to 5,925 MHz. An amendment to the carrier sense multiple access/collision avoidance (CSMA/CA) standard is foreseen fitting wireless access in vehicular environments (WAVE) and supporting intelligent transportation systems (ITS) applications. The architecture of communication in ITS (ITSC) is specified by the standard ETSI EN 302 665. The open systems interconnection (OSI) basic reference model (ISO/IEC 7498-1 standard) is followed to propose an ITSC open system architecture and its layered functionalities.

Generalization to multiple wireless technologies has been sought within the Communications Access for Land Mobiles (CALM) program, covering 2G and 3G cellular, Wi-Fi, DSRC, and also the millimetric (63–64 GHz) band for radar communication and transponders. ITS subsystems are identified to be vehicle, roadside, central, and personal (handheld devices) ITS stations enabling communication using any of the available communication technologies. Vehicle, roadside, or central stations are constituted by the functionalities embedded in ITS-S host, ITS-S gateway, ITS-S router, and ITS-S border router. A personal ITS station is used for mobile phones or personal digital assistants (PDA), offering interface functionality [23].

7.6   Potential Applications in an Autonomous Vehicle

There are many possible applications for vehicle-to-vehicle and vehicle-to-infra-structure communication in an autonomous vehicle, including:

- Intention notification among vehicles;
- Platooning and merging coordination;
- Intersection negotiation;
- Sensor sharing and multivehicle fusion and situation awareness;
- Map updating.

In this section we will develop a few examples. Two important operations in highway driving will be presented: platoons [24] and adaptive cruise control (ACC), and the control of merging traffic. A theoretical analysis of an ACC policy is given to demonstrate the performance improvement with intervehicle communication. We simulate collision avoidance and merging scenarios to illustrate the effectiveness of wireless communication in intelligent vehicle applications and to demonstrate the technology. Merging followed by an emergent braking scenario is studied to further show the effect of wireless communication on the improvement of possible automated and cooperative driving [21].

For urban driving we present a stop-and-go control system with and without wireless communication. In the stop-and-go scenario, decelerations and emergency braking to avoid collisions is controlled by an automated driving system with a given sensing range, while starting or acceleration is executed by the human driver.

In the simulations that follow we use a simplified first order longitudinal model of the vehicle with a time constant of 10 seconds and a unity gain. Dimensions and acceleration limits for several classes of vehicles are shown in Table 7.3. We will only evaluate the effect of the loop closure delay (caused by communications delays) on safety. The details of how the delay is produced will not be considered. Thus, there is no need to precisely model the physical layer or MAC layer of the V2V communication devices.

Modeling driver behavior is a complex but necessary part of simulations used for studying new technologies since individual drivers perceive and react differently. The driver's perception and reaction time plays a significant role in all problems related to safety. A simple model, such as a parameterized first- or second-order linear system, can be used to define a set of driver types, and instances of each type can be distributed among the simulated vehicles. The driver model determines the driver's acceleration, deceleration, and response to warnings. Thus different drivers may take different actions when they face the same situations. For example,

Table 7.3   Physical Vehicle Characteristics

Type	Width	Length	Acceleration	Deceleration
Passenger car	2.13m	5.79m	0.2g m/s^2	0.31g m/s^2
Truck	2.59m	9.14m	0.2g m/s^2	0.31g m/s^2
Unit bus	2.59m	12.19m	0.2g m/s^2	0.31g m/s^2
Motorcycle	0.76m	2.13m	0.2g m/s^2	0.31g m/s^2

when drivers notice a possible collision ahead, conservative drivers may perform an emergency braking action while more aggressive drivers will only release the throttle pedal.

7.6.1   Platoons and Adaptive Cruise Control (ACC)

Cruise control systems regulate vehicle speed to maintain a set point when there is no vehicle or obstacle ahead. In an ACC system, when a slower vehicle is detected the ACC controlled vehicle will follow the vehicle at a safe distance by adjusting its speed. A convoy or platoon can be established with a set of vehicles using ACC.

The most conventional ACC algorithm is constant time gap control. Consider two cars (leader and follower), shown in Figure 7.11, with lateral positions of X_L and X_F, respectively.

A driver maintains speed based on the perceived gap to the vehicle ahead and the speed differential. Thus, if we consider a very simplistic point-mass model for a car the dynamics would be:

$$\ddot{x}_F = u(t)$$

where $u(t)$ is the commanded input from the driver to the ACC system. Then $u(t)$ could be selected as

$$u(t) = k_1\left(x_L - x_F\right) + k_2\left(\dot{x}_L - \dot{x}_F\right) + k_3$$

where k_1, k_2, k_3 represent PD and feedforward controller gains that might be picked subconsciously by a driver or after analysis by the ACC designer. Another way of formulating the same problem is

$$\begin{cases} u(t) = -1\big/ h\left(\dot{\varepsilon} + \lambda\delta\right) \\ \varepsilon = x_L - x_F + L \\ \delta = x_L - x_F + L + h\dot{x}_L \end{cases} \tag{7.1}$$

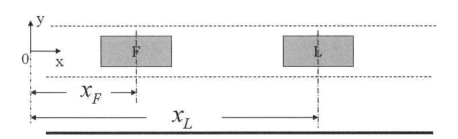

Figure 7.11   ACC algorithm.

where

> L: Desired constant spacing between vehicles (typically 2–10m);
>
> ε: Spacing errors;
>
> δ: Spacing errors including time headway control;
>
> h: Constant time gap (CTG), usually between 1 and 2 seconds;
>
> λ: control gain, determines the convergence rate of spacing error δ.

The three parameters (h, λ, L) are selected by the designer. It has been shown that this control law can ensure string stability [25–27]. If we are interested in platoon formation, this control can be applied to every vehicle in a platoon.

The ACC policy in a highway scenario in which eight vehicles are grouped in a platoon is simulated in the following examples. A typical ACC controlled car-following case is chosen [25–28] with a desired constant time headway of 1 second, a constant desired spacing between vehicles in a platoon of 5m, and a convergence rate of 0.2. The effect of intervehicle communication is simulated. Extensive studies have been carried out by researchers at PATH with similar results [29].

7.6.1.1 A Platoon Example

An ACC policy without intervehicle communication can be studied using a traffic simulator. To simulate car-following on a highway, all the vehicles are assumed to have an initial speed of 60 mph. Their time gaps are assigned as 1.5 seconds and safety distance L is chosen as 5m, which is also the sensing range. A stationary obstacle, which might be a stopped vehicle, a lane closure due to construction, or cargo dropped on the road, is placed into the lane of the oncoming vehicles. No communications exist in the platoon, which means a vehicle can only see another vehicle ahead when the leader is in its detection range.

Figure 7.12(a) shows the position profile of the first three vehicles in the simulated platoon. It shows that when the first vehicle of the platoon senses the stationary obstacle it slows down and all the followed vehicles begin to slow down to attempt to avoid a collision. However, due to the zero velocity of the obstacle and small sensing range of the vehicles, collisions occur. The scenario with intervehicle communication is also simulated. When the first vehicle of the platoon senses the obstacle, it sends out the warning message immediately and all the following vehicles in the platoon are able to decelerate simultaneously. Figure 7.12(b) shows that although the lead vehicle still collides with the obstacle due to the short sensing range, the desired constant time gap (CTG) is well maintained between other vehicles due to the communication and early reaction capabilities.

The speed profile of the first three vehicles using ACC without communications for the same scenario is plotted in Figure 7.13(a). The longitudinal velocity of vehicles decreases successively as the detection of the front vehicle's deceleration gradually propagates through the platoon. The speed profiles of the vehicles in an intervehicle communication system are plotted in Figure 7.13(b). As the warning message is received by all the vehicles at almost same time (with only small delays due to communication), all the following vehicles slow down simultaneously and avoid possible collisions. Communication also allows us to avoid the "slinky

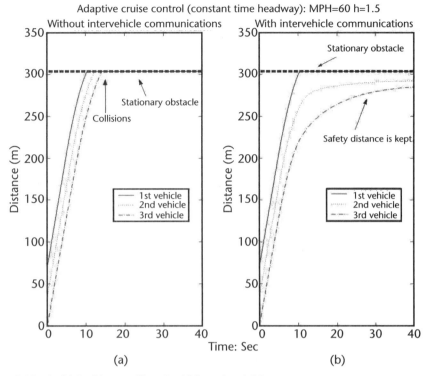

Figure 7.12   (a, b) Position profiles of vehicles using ACC.

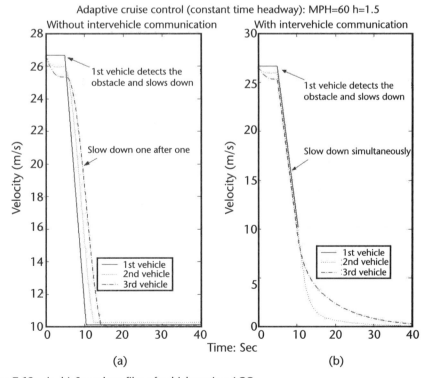

Figure 7.13   (a, b) Speed profiles of vehicles using ACC.

effect," in which small perturbations are amplified as they travel back through the platoon leading to oscillations of increasing amplitude and eventually to instability and collisions.

7.6.1.2 Analysis of Platoon Performance

We can study the effects of intervehicle communication based on the mathematic model as shown in (7.1), namely,

$$\ddot{x}_{ides} = -\frac{1}{h}\left(\dot{\varepsilon}_i + \lambda\,\dot{\delta}_i\right)$$

Expanding for ε_i and δ_i, we obtain

$$\ddot{x}_{ides} = -\frac{1}{h}\left(\dot{x}_i - \dot{x}_{i-1} + \lambda\left(x_i - x_{i-1} + L + h\,\dot{x}_i\right)\right) = -\frac{1}{h}\left(\Delta v + \lambda\left(S + h\,\dot{x}_i\right)\right)$$

where $\Delta v = \dot{x}_i - \dot{x}_{i-1}$ are the relative velocity changes between vehicles in a certain information updating time interval, $S = x_i - x_{i-1} + L$ is the longitudinal position difference between vehicles, and L is the safety distance required between vehicles. Note that S is usually a negative value and its absolute value cannot exceed L since the ith vehicle follows the $(i-1)$th vehicle.

Thus the relationships between velocity, potential velocity changes, and information distance are given by

$$\frac{-h\bar{a} - \Delta v - \lambda S}{\lambda h} \le \dot{x}_i \le \frac{-h\underline{a} - \Delta v - \lambda S}{\lambda h} \tag{7.2}$$

Figure 7.14 shows both the upper limit and the lower limit of (7.2) An automated vehicle has a certain time interval to exchange information and then take action accordingly. The duration of this time interval depends on system parameters such as onboard system data processing time, vehicle reaction time, and communication interval (if intervehicle communication is available). Then Δv represents the relative velocity changes between two vehicles during this time interval. If we consider a platoon on the highway in which all the vehicles have reached steady state (i.e., maintain the same speed as the lead vehicle), then in a given time interval Δt the maximum relative velocity changes is $\Delta v = |(\bar{a} - \underline{a}) \times \Delta t|$, which occurs in the worst-case scenario of the leading vehicle performing an emergency deceleration while the following vehicle is accelerating at its maximum rate. The information distance is defined as the longitudinal position difference between these two vehicles S at a given time stamp and can be viewed as the distance that is needed for information exchanges.

We simulated two cases in which Δt is 0.1 and 1.0 second, corresponding to a Δv of 0.5 and 5.0 m/s, respectively. The results are plotted in Figure 7.14. The area between the upper limit and lower limit lines is the area of safe controllability

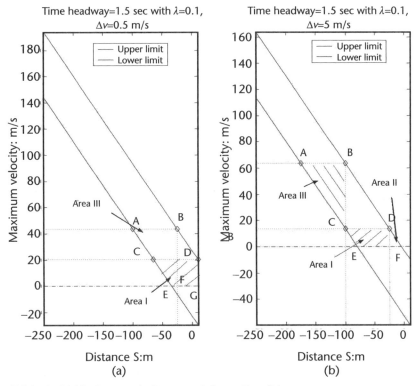

Figure 7.14   (a, b) Maximum velocity versus information distance.

for velocity according to the specific information distance. For example, in Figure 7.14(a), line BF indicates that when the information distance is 25.03m, the vehicle is safely controllable when its velocity is between 0 m/s and 43.375 m/s. The area above the upper limit is an uncontrollable region. Line AB, on the other hand, indicates that a vehicle with the velocity of 43.375 m/s requires an information distance of at least 25.03m and also that no action needs to be taken when the distance between two vehicles are greater than 100m. In both figures, area I represents the requirement on the information distance for a given velocity while area III represents the safely controllable range of vehicle velocity for the given information distance.

Comparisons between Figure 7.14(a) and Figure 7.14(b) show that the required information distance can be shortened by reducing the time interval. However, it is impractical to reduce the time interval below 0.1 second due to the equipment constraints. On the other hand, increasing information distance by means of intervehicle communication can ease the system requirement. As shown in Figure 7.14(b), line BC indicates that, even with a long update interval of 1 second, the vehicle's velocity can reach 63.727 m/s with an information distance of 100m. The results also show that vehicle needs to take action only when it falls into the area between the two lines.

This analysis shows that with intervehicle communications, the updating time interval for the control of the vehicle can be larger; in other words, we do not need a communication system with a short transmission interval to ensure the controllability of vehicles. This eases the requirement on both the onboard processing system and the communication equipment bandwidth.

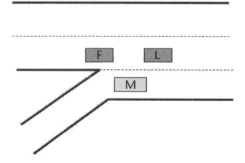

Figure 7.15   Ramp merge problem.

7.6.2   Merging Traffic

Another important operation in highway systems is the control of merging traffic. A good control system can avoid possible collisions and provide smooth operation of the roadway, which minimize delays, traffic jams, and collisions caused by the merging vehicle. This is also an application where vehicle-to-vehicle communication can improve safety and performance.

Many theoretical and simulation studies have been conducted for the merging problem [30–37]. Consider the three vehicles shown in Figure 7.15 that will be directly affected when the vehicle on the entrance ramp attempts to merge. Vehicle M needs to maintain a certain safety distance from vehicle F and vehicle L so that it can perform a collision-free and smooth merge operation, but a problem arises when there is insufficient space for vehicle M to merge between vehicle L and vehicle F. Vehicle M therefore needs to slow down or stop to wait for an appropriate gap and avoid a collision. In reality a wait is unnecessary if there is an open slot ahead of vehicle L or in another lane. A merge control algorithm is thereby needed to solve this problem by adjusting vehicle M's approach speed to reach an open slot for merging. To implement the proposed merge control algorithm, intervehicle communication is required to exchange information such as position and speed between vehicles.

7.6.2.1   A Merging Control Algorithm

Without the aid of intervehicle communication, vehicle M can only merge into the main lane after it detects a sufficient gap using some onboard sensor. Neither vehicle F nor vehicle L will cooperate until vehicle M enters their sensing range, at which time vehicle F may be obliged to rapidly slow down to create a suitable gap.

With the aid of intervehicle communication, vehicle M broadcasts its position and speed at some point on the onramp. A virtual mapping method as shown in Figure 7.16 may be employed to map the merging vehicle onto the main lane. Therefore, the critical issue in the merging control algorithm considered is to generate enough spacing between vehicle F and virtual vehicle M as well as virtual vehicle M and vehicle L [31, 32]. The key idea of this simple algorithm is to best utilize the open spot ahead of vehicle L since it is within the communication range. Using IVC, vehicles may exchange information and make cooperative maneuvers to move the open spot backward and accommodate vehicle M.

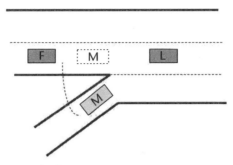

Figure 7.16   Virtual mapping method.

Figure 7.17 shows the detailed merging control algorithm. Upon entering the system, the merging vehicle broadcasts its position, time stamp, speed, acceleration rate, heading, and size [30]. Vehicles within communication range broadcast their own position, speed, headway, and size to the merging vehicle. After receiving the information, the merging vehicle virtually maps itself onto the main lanes and looks forward to find the closest open spot from itself at the predicted time it will reach the merging point. It also calculates whether there is any road contention between itself and its closest follower. It will then broadcast the position of the closest open spot, which is ahead of itself and the closest follower's ID if there is a potential road contention. Based on this information, the vehicles on the main lanes compute whether they are affected and determine what to do. Those vehicles that are ahead of the merging vehicle and behind the closest open spot will change lane and/or speed up, depending on whether the open spot is in the rightmost lane. The open spot is then effectively moved backward and matched to the position of the merging vehicle. If there is a potential road contention between the closest follower and the merging vehicle, the closest follower will slow down until the required open spacing is achieved. It should be mentioned that the merging vehicle does not look for the closest open spot, which is behind itself, since this would create a slow down in the main lane. It also should be noted that all the affected leaders

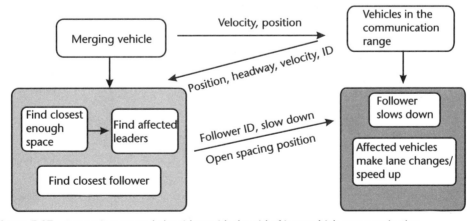

Figure 7.17   A merging control algorithm with the aid of intervehicle communication.

will match their speeds to the merging vehicle until the merging vehicle completes its lane change.

In the following simulations we assume a typical ACC controlled car-following case with desired constant time headway of 1 second, constant spacing between vehicles in a platoon of 5 meters, and convergence rate of 0.2 on the main lane.

7.6.2.2 Merging Example

One possibility for merging is that an affected vehicle changes lanes to accommodate the merging vehicle. We omit simulations for this case. Instead, we present the case in which vehicles alter their speed to accommodate the merging vehicle. This may be the case when there is no vacancy in the adjacent lane of traffic or no other traffic lane exists.

Figure 7.18 shows a typical speed changing scenario in which two affected vehicles are in a platoon with a spacing less than the threshold and in which the ideal merging point for the merging vehicle is between these two affected vehicles. At the 24.76-second time point, both the two affected vehicles receive the broadcast from the merging vehicle and determine the road contention. The leading vehicle starts to accelerate while the following vehicle starts to decelerate to accommodate the merging vehicle as shown in Figure 7.18. After the spacing between these two vehicles exceeds the threshold, both vehicles revert to the set point speed. Figure 7.18(c) shows the changes in intervehicle distances within the platoon.

Without intervehicle communication, merging vehicles have to slow down or even stop when the merging point is occupied by another vehicle; therefore it takes

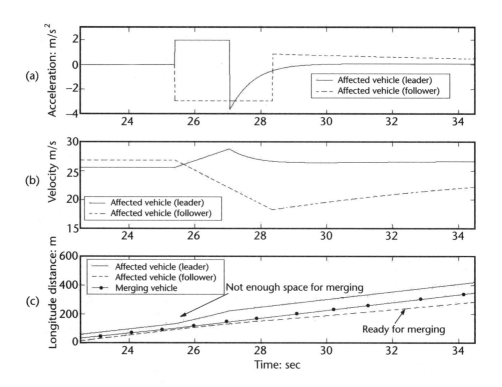

Figure 7.18 (a–c) Merging by altering vehicle speeds.

longer time for vehicles to merge into the right lane, which results in delay. Simulations show the effect of IVC on the delay due to the occupied merging point. Cases with traffic throughput of 1,200, 1,800, 2,400, and 3,200 vehicles/hour in the right lane are simulated. The delay of the merging vehicle is calculated by subtracting the time required for vehicles to merge when there is no traffic in the main lane from the time obtained by simulation with other traffic. For each throughput rate 100 merging vehicles are simulated. The average delays under various conditions are plotted in Figure 7.19. The data shows that as traffic throughput increases from 1,200 to 3,200 vehicles/hour, the average delay is increased from 0.3 second to 1.5 seconds when no IVC is employed because the probability of vehicles arriving at the merging point simultaneously increases. However, the delay is kept almost constant around 0 second when IVC is employed.

Algorithms such as this one also assist in avoiding collisions due to unsafe manual merging maneuvers.

7.6.3   Urban Driving with Stop-and-Go Traffic

Stop-and-go operation is a common situation in an urban area. In this example we consider a restrictive version of stop-and-go. We assume that vehicle deceleration and stopping for time-critical situations will be accomplished by an automated system (adaptive cruise control) in an acceptable sensing range while the starting and acceleration of the vehicle will be controlled by the human driver. Such a scenario

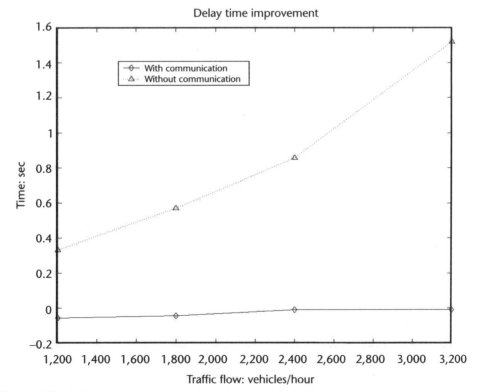

Figure 7.19   Delay time improvements.

may be appropriate when the traffic flow consists of both automated and manually controlled vehicles [38].

The stop-and-go in urban area simulation scenario is studied by simulating 12 vehicles, which are divided in two groups. The effect of intervehicle communication is simulated. A simple human driver model was assumed to represent the reactions of the driver for starting and acceleration control. All the vehicles are initially traveling with a velocity of 25 mph with time gaps of 1.5 seconds and a safety distance L chosen as 5m, which is also the sensing range. A stationary obstacle is placed into the lane of the oncoming vehicles.

7.6.3.1 Example of Two Groups Without Intervehicle Communication

Since no intervehicle communication is available, all the following vehicles have only the information (distance and velocity) of the vehicle directly ahead, and the vehicle can only see the vehicle ahead when the leader is within its sensing range. No information can be exchanged between two groups.

Figure 7.20(a) plots the trace files of the two groups of vehicles showing the spacing between leading and following vehicles. When the stationary obstacle is sensed by the leading vehicle in the first group, all the following vehicles within the sensing range of the front vehicle decelerate to avoid collision. Hence, spacing between the vehicles is reduced to a value that barely avoids a collision. Another observation is that the two groups merge. In this case the time gap for each ve-

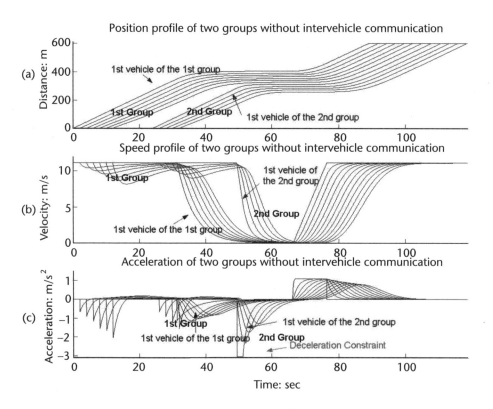

Figure 7.20 (a–c) The stop-and-go scenario without communication.

hicle is large enough to safely stop and avoid the stationary obstacle during this low-speed operation.

The speed profiles of vehicles without communications in this simulation scenario are shown in Figure 7.20(b). The longitudinal velocities of the vehicles in the groups decrease by following the leading vehicle's deceleration. Since there is no communication between the vehicles or the groups, the speed profile of the vehicles in the platoons is activated only by the leading vehicle's deceleration. Acceleration/ deceleration profiles of the vehicles are given in Figure 7.20(c). Since the time gap in this example is large enough for stop-and-go or collision avoidance by emergency braking, the emergency deceleration limit of 0.31g is sufficient to stop the leader vehicle of the second group within its 5-meter sensing distance.

7.6.3.2   Example of Two Groups with Intervehicle Communication

In this scenario, all the vehicles are assumed to be able to communicate, receive the information from the first vehicle of the first group, and perform cooperative driving with the aid of intervehicle communication. The position profiles of the vehicles are plotted in Figure 7.21(a). When the first vehicle of the first group senses an obstacle ahead it decelerates to avoid a collision and sends out a warning message. All the following vehicles are informed of the possible hazard at essentially the same time. The desired time gaps are well maintained between vehicles due to communication and early reaction capabilities.

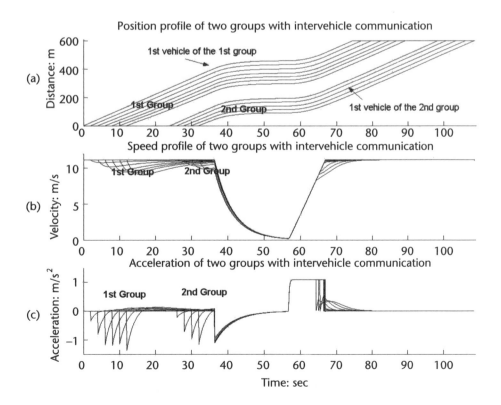

Figure 7.21   Stop-and-go scenario with communication.

The speed profiles of the vehicles within both groups are shown in Figure 7.21(b) and the acceleration/deceleration profiles of the vehicles in both groups are shown in Figure 7.21(c). Since the warning message is received by all the vehicles in both groups, the vehicles start braking simultaneously in order to avoid possible obstacles and collisions and maintain a safe following distance or assigned constant time gap profiles; smooth deceleration and acceleration profiles are obtained.

References

[1]   http://www.its.dot.gov.

[2]   http://www.vehicle-infrastructure.org.

[3]   http://www.esafetysupport.org/en/esafety_activities/.

[4]   http://ec.europa.eu/transport/wcm/road_safety/erso/knowledge/Content/04_esave/esafety.htm.

[5]   "Car2Car Communication Consortium Manifesto," http://www.car-to-car.org/fileadmin/downloads/C2C-CC_manifesto_v1.1.pdf.

[6]   http://www.comesafety.org.

[7]   http://www.cvisproject.org.

[8]   http://www.esafetysupport.org/en/esafety_activities/related_projects/index.html.

[9]   http://www.ertico.com.

[10]   Tsugawa, S., "A History of Automated Highway Systems in Japan and Future Issues," *Proc. IEEE Conference on Vehicular Electronics and Safety*, Columbus, OH, September 22–24, 2008.

[11]   Tsugawa, S., "Energy ITS: Another Application of Vehicular Communications," *IEEE Communications Magazine*, November 2010, pp. 120–126.

[12]   http://www.mlit.go.jp/road/ITS/.

[13]   Korkmaz, G., E. Ekici, and F. Özguner, "An Efficient Ad-Hoc Multi-Hop Broadcast Protocol for Inter-Vehicle Communication Systems," *IEEE International Conference on Communications (ICC 2006)*, 2006.

[14]   http://www.its.dot.gov/cicas/.

[15]   http://www.vics.or.jp/english/vics/pdf/vics_pamph.pdf.

[16]   Krogmeier, J. V., and N. B. Shroff, *Final Report: Wireless Local Area Network for ITS Communications Using the 220 MHz ITS Spectral Allocation*, FHWA/IN/JTRP-99/12, April 2000.

[17]   Fitz, M. P., et al., "A 220 MHz Narrowband Wireless Testbed for ITS Applications," *The Fourth International Symposium on Wireless Personal Multimedia Communications*, Aalborg, Denmark, September 2001.

[18]   http://www.ntia.doc.gov/osmhome/spectrumreform/Spectrum_Plans_2007/Transportation_Strategic_Spectrum_Plan_Nov2007.pdf.

[19]   Tokuda, K., M. Akiyama, and H. Fujii, "DOLPHIN for Inter-Vehicle Communications System," *Proceedings of the IEEE Intelligent Vehicle Symposium*, 2000, pp. 504–509.

[20]   Shiraki, Y., et al., "Development of an Inter-Vehicle Communications System," *OKI Technical Review 187*, Vol. 68, September 2001, pp. 11–13, http://www.oki.com/en/otr/downloads/otr-187-05.pdf.

[21]   Tsugawa, S., et al., "A Cooperative Driving System with Automated Vehicles and Inter-Vehicle Communications in Demo 2000," *Proc. IEEE Conference on Intelligent Transportation Systems*, Oakland, CA, 2001, pp. 918–923.

[22]   DRAFT SAE J2735 Dedicated Short Range Communications (DSRC) Message Set Dictionary Rev 29, SAE International, http://www.itsware.net/ITSschemas/DSRC/.

[23]   http://www.calm.hu.

[24]   Shladover, S. E., et al., "Automated Vehicle Control Development in the PATH Program,"
        IEEE Transactions on Vehicular Technology, Vol. 40, 1991, pp. 114–130.

[25]   Swaroop, D., and K. R. Rajagopal, "Intelligent Cruise Control Systems and Traffic
        Flow Stability," *Transportation Research Part C: Emerging Technologies*, Vol. 7, 1999,
        pp. 329–352.

[26]   Rajamani, R., and C. Zhu, "Semi-Autonomous Adaptive Cruise Control Systems," *Proc.
        Conf. on American Control*, San Diego, CA, 1999, pp. 1491–1495.

[27]   Rajamani, R., and C. Zhu, "Semi-Autonomous Adaptive Cruise Control Systems," *IEEE
        Transactions on Vehicular Technology*, Vol. 51, 2002, pp. 1186–1192.

[28]   Zhou, J., and H. Peng, "Range Policy of Adaptive Cruise Control Vehicles for Improved
        Flow Stability and String Stability," *IEEE Transactions on Intelligent Transportation Sys-
        tems*, Vol. 6, 2005, pp. 229–237.

[29]   Xu, Q., et al., "Vehicle-to-Vehicle Safety Messaging in DSRC," *Proc. First ACM Workshop
        on Vehicular Ad Hoc Networks (VANET 2004)*, Philadelphia, PA, 2004, pp. 19–28.

[30]   *Vehicle Safety Communications Project Task 3 Final Report—Identify Intelligent Vehicle
        Safety Applications Enabled by DSRC*, U.S. DOT HS 809 859 (NHTSA), March 2005,
        http://www.nhtsa.gov/DOT/NHTSA/NRD/Multimedia/PDFs/Crash%20Avoidance/2005/
        CAMP3scr.pdf.

[31]   Sakaguchi, T., A. Uno, and S. Tsugawa, "Inter-Vehicle Communications for Merging Con-
        trol," *Proc. IEEE Conf. on Vehicle Electronics*, Piscataway, NJ, 1999, pp. 365–370.

[32]   Uno, A., T. Sakaguchi, and S. Tsugawa, "A Merging Control Algorithm Based on Inter-
        Vehicle Communication," *Proc. IEEE Conf. on Intelligent Transportation Systems*, Tokyo,
        Japan, 1999, pp. 783–787.

[33]   Fenton, R. E., and P. M. Chu, "On Vehicle Automatic Longitudinal Control," *Transporta-
        tion Science*, Vol. 11, 1977, pp. 73–91.

[34]   Fenton, R. E., "IVHS/AHS: Driving into the Future," *IEEE Control Systems Magazine*,
        Vol. 14, 1994, pp. 13–20.

[35]   Ioannou, P. A., and M. Stefanovic, "Evaluation of ACC Vehicles in Mixed Traffic: Lane
        Change Effects and Sensitivity Analysis," *IEEE Transactions on Intelligent Transportation
        Systems*, Vol. 6, 2005, pp. 79–89.

[36]   Drew, D. R., *Traffic Flow Theory and Control*, New York: McGraw-Hill, 1968.

[37]   Takasaki, G. M., and R. E. Fenton, "On the Identification of Vehicle Longitudinal Dynam-
        ics," *IEEE Transactions on Automatic Control*, Vol. 22, 1977, pp. 610–615.

[38]   Acarman, T., Y. Liu, and Ü. Özgüner, "Intelligent Cruise Control Stop and Go with and
        Without Communication," *Proc. Conf. on American Control*, Minneapolis, MN, 2006,
        pp. 4356–4361.

Selected Bibliography

ASTM E2158-01 Standard Specification for Dedicated Short Range Communication (DSRC)
Physical Layer Using Microwave in the 902 to 928 MHz Band.

ETSI ES 202 663 (V1.1.1), "Intelligent Transport System (ITS); European Profile Standard for
the Physical and Medium Access Control Layer of Intelligent Transport Systems Operating in the
5 GHz Frequency Band," 2010.

ETSI EN 301 893 (V1.5.1), "Broadband Radio Access Networks (BRAN); 5 GHz High Performance RLAN; Harmonized EN Covering the Essential Requirements of Article 3.2 of the
R&TTE Directive."

ETSI EN 302 571 (V1.1.1), "Intelligent Transport Systems (ITS); Radiocommunications Equipment Operating in the 5.855 MHz to 5.925 MHz Frequency Band; Harmonized EN Covering the Essential Requirements of Article 3.2 of the R&TTE Directive."

ETSI EN 302 665 (v1.1.1), "Intelligent Transport Systems (ITS); Communications Architecture," 2010.

ETSI TS 102 665 (V1.1.1), "Intelligent Transport Systems (ITS); Vehicular Communications; Architecture."

ETSI TS 102 687 (V1.1.1), "Intelligent Transport Systems (ITS); Transmitter Power Control Mechanism for Intelligent Transport Systems Operating in the 5 GHz Range."

The Institute of Electrical and Electronics Engineers (IEEE), "Wireless LAN Medium Access Control (MAC) and Physical Layer (PHY) Specifications," http://standards.ieee.org, ANSI/IEEE Std.802.11, 1999 (also know as ISO/IEC 8802-11:1999(E), 2007).

The Institute of Electrical and Electronics Engineers (IEEE), "IEEE 802.11e/D4.4, Supplement to Part 11: Wireless Medium Access Control (MAC) and Physical Layer (PHY) Specifications: Medium Access Control (MAC) Enhancements for Quality of Service QoS," June 2003.

IEEE P802.11k: "IEEE Standard for Information Technology—Telecommunications and Information Exchange Between Systems—Local and Metropolitan Area Networks—Specific Requirements Part 11: Wireless LAN Medium Access Control (MAC) and Physical Layer (PHY) Specifications Amendment 1: Radio Resource Measurement of Wireless LANs," 2008.

IEEE P802.11pTM/D8.0: "Draft Standard for Information Technology—Telecommunications and Information Exchange Between Systems—Local and Metropolitan Area Networks—Specific Requirements—Part 11: Wireless LAN Medium Access Control (MAC) and Physical Layer (PHY) Specifications; Amendment 7: Wireless Access in Vehicular Environments," 2009.

"IEEE Standard for Information Technology—Telecommunications and Information Exchange Between Systems—Local and Metropolitan Area Networks—Specific Requirements Part II: Wireless LAN Medium Access Control (MAC) and Physical Layer (PHY) Specifications," IEEE Std. 802.11g-2003 (Amendment to IEEE Std. 802.11, 1999 Edn. (Reaff 2003) as amended by IEEE Stds. 802.11a-1999, 802.11b-1999, 802.11b-1999/Cor 1-2001, and 802.11d-2001), 2003.

"IEEE Std. 1609.1—2006 IEEE Trial-Use Standard for Wireless Access in Vehicular Environments (WAVE)—Resource Manager," IEEE Std. 1609.1-2006, 2006.

"IEEE Trial-Use Standard for Wireless Access in Vehicular Environments—Security Services for Applications and Management Messages," IEEE Std. 1609.2-2006, 2006.

"IEEE Trial-Use Standard for Wireless Access in Vehicular Environments (WAVE)—Multi-Channel Operation," IEEE Std. 1609.4-2006, 2006.

"IEEE Trial-Use Standard for Wireless Access in Vehicular Environments (WAVE)—Networking Services," IEEE Std. 1609.3-2007, April 20, 2007.

ISO/IEC 7498-1, "Information Technology—Open Systems Interconnection—Basic Reference Model: The Basic Model," 1994.

CHAPTER 8

Conclusions

8.1 Some Related Problems

In this concluding chapter of the book we want to mention two problems in the development of autonomous vehicles. We picked these two as still open for research and worth investigating, and we also view them as immediate contributors.

8.1.1 Fault Tolerance

Fault tolerance in an autonomous system relies on the quantity and quality of environmental information available. The amount of reliable, real-time data able to be extracted from the surroundings is the limiting factor, not the control systems. Another obstacle to the actual implementation of intelligent vehicle systems onto public roadways is the public acceptance of such technologies as being safe and reliable.

Despite the current concerns and skepticism over the deployment of intelligent vehicle systems, the possibility of increased automobile density on the highways and the need for increased safety and reduced pollution all help support further development of intelligent vehicle (IV) systems. Leading areas for development are in sensor fusion, validation and fault detection, isolation, and tolerance. These areas of research will help provide the safety and robustness required for IV systems to gain public acceptance and real-world implementation.

In two papers in 1998 [1, 2] we proposed a framework for addressing these issues: the Sensor Fault Controller Scheme (SFCS). This framework is applied to the problems of advanced cruise control (ACC) and lane keeping (LK) with examples using the automated vehicles developed at The Ohio State University. The proposed SFCS framework is comprised of four tiers, each containing individual modules that pertain to separate tasks. The complete SFCS framework is itself modular, providing a closed-loop control path around the sensors and providing a concise, common data type to all high-level control and computation routines. The four separate tiers in the SFCS framework are the physical sensor, data preprocessing, data processing, and control and computation (Figure 8.1). The physical sensor tier is where the actual sensors and their associated control algorithms reside. The next phase in the SFCS framework is the data preprocessing tier, where the raw sensor

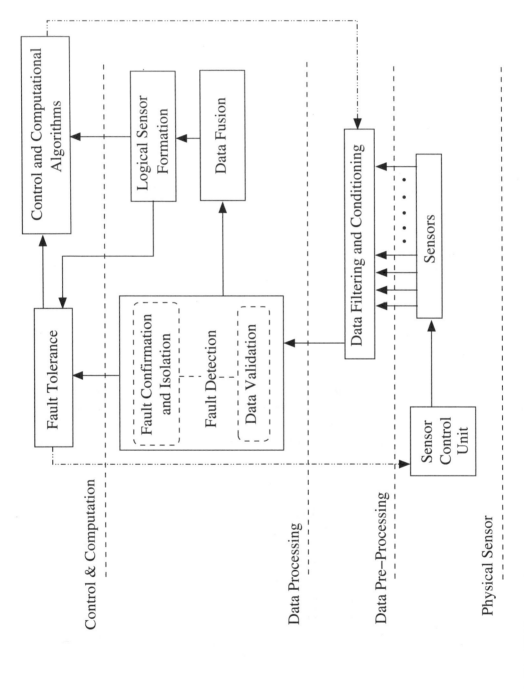

Figure 8.1   The SFCS framework.

data is passed through initial filtering and conditioning algorithms. These processes look for outlying data points using rule bases. This filtered data is then transmitted to the data processing tier where the data validation, fault detection and isolation, and data fusion algorithms are located. The sensor data is packaged into logical sensor packets for transmission to the control and computational tier at this level as well. The highest phase in the SFCS framework, the control and computational tier is a terminal for data computation. This plane is where the existent IV system control algorithms reside, as well as, the fault tolerance decision module. Further analysis of the four tiers, their associated modules, and specific data transmission protocols are discussed in the papers [1, 2].

Certain guidelines have to be adopted to specify what is meant by ACC and LK for this approach, and to likewise dictate the sensor data that is of interest and in what regions. ACC is concerned with the detection of all obstacles in the forward path of motion of the home vehicle with relative speeds of 10 m/s and within a look-ahead distance of 10m to 100m. The main specification for the LK system is the acceptable offset distance—how far the vehicle is allowed to drift from the center of the lane. The lateral offset of the vehicle is desired to be regulated to zero; however, given the resolution of the lateral sensing devices, a 5-cm offset distance could be acceptable for this research. The offset rate is also of interest in the validation of sensor measurements and for control purposes.

We provide SFCS as an example of a possible approach. The readers are encouraged to develop other possible approaches [3].

8.1.2 Driver Modeling

Driver modeling is especially important for at least three reasons:

- Developing autonomous vehicle decision logic to emulate humans;
- Understanding (simulating) how humans will behave in mixed traffic with autonomous vehicles;
- Developing safety technologies that could immediately be used in intelligent vehicles.

It is therefore desirable to integrate a comprehensive, quantitative human driver model with automation into micro-level simulation tools for traffic simulation and performance evaluation at the traffic level because; this is because humans not only bring a number of skills for driving targets that most artificial intelligent technology can not accurately simulate, but also help to improve driving safety and driving learning. In this section, based on our work reported first in [4, 5] and further developed in [6], we construct a human driver model and examine the driver behaviors at both signalized and unsignalized intersections.

Since the 1960s, the human driver model has attracted the attention of researchers in transportation studies. Descriptive models, which provide straightforward ideas of human driver behavior under different kinds of scenarios, were initially developed. Risk-based human driver models combine the driver's motivation, experience, and so forth with risk perception, acceptance, and so forth and focus on the driver's psychological thinking process. These types of models are also used for driver assistance systems design and development. The cognitive human driver model has attracted researchers' attention for several years. The essential quality of the cognitive human driver model focuses on human drivers' psychological activities during the driving. Cognitive models on the other hand can help to develop understanding of driver behavior. The cognitive simulation model of the driver (COSMODRIVE) was developed at the French Institute for Transportation Research [7]. PATH researchers extended and organized the COSMODRIVE framework for the purpose of driver modeling in their SmartAHS [8–11]. The model allowed simultaneous simulations of vehicles controlled by drivers and semiautomated systems for comparisons. A driver's knowledge database and the cognitive process underlying the driving activity contribute to the cognitive approaches.

We first present a general architecture of the human driver model as shown in Figure 8.2. This model is based on the structure of COSMODRIVE. The model consists of seven modules, which can be divided into two groups: the external and internal view of the driver. The environment module is the external to the group, while all other modules are in the internal group.

- *Environment module*: The environment module represents the world outside the vehicle. It takes traffic environment (traffic flow, traffic composition, and so forth), road environment (speed limit, lane numbers, and so forth), and weather/visibility (sunny/rainy/windy, good/poor, and so forth) into account.

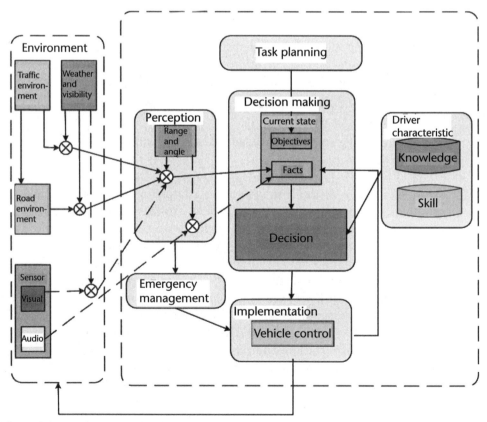

Figure 8.2 Configuration of a driver model.

Additional sensors, such as camera/LIDAR may also provide the driver with necessary data on the environment. Once a driver-assistance system is available, the system assistant message will also be served as a complement to the environment.

- *Perception module:* The perception module represents visual and audio sensing. The data generated by the perception module includes estimation of velocity, direction, distance between vehicles, and so forth in range and angle. In a microscopic traffic simulator, when considering the changes of visibility for the drivers, we can simply adjust the parameters of range and angle according to the situation in the perception module. In the real world, range and angle are based on the driving environment, for example, blizzard weather leads to short range and small angle.

- *Task-planning module:* The task-panning module provides the decision-making module with information on which direction the vehicle is going.

- *Driver characteristics module:* The essential function of the driver characteristics module is to predict a human driver's psychological activities based on his/her knowledge, which contains both driving knowledge-based and traffic rule–based information, and his/her driving skill, which indicates his/her ability of driving (novice/expert). It changes as the subject driver changes.

- *Decision-making module:* The decision-making module is the most important part of the driver model. It acts as a higher-level controller for the vehicle. The decision is made based on the perceptive information, its itinerary from task planning module, the driver's own characteristics, and the vehicle's current state.

- *Implementation module:* The implementation module is responsible for the two- dimensional control of the vehicle based on the information it receives from the decision-making module.

- *Emergency management module:* The emergency management module deals with unexpected/irregular emergency, such as another vehicle's traffic violation and obstacle avoidance.

The reader is encouraged to compare this to the functional architecture presented in Chapter 3.

Consider the multilane intersection shown in Figure 8.3 [4, 5], which provides a number of examples of finite state machines representing the decision-making module indicated above. Figure 8.4 is an example. A probabilistic version of such models of human driving is provided in [6] where it is used to estimate what the driver intent could be (e.g., running a red light).

8.2   And the Beat Goes On

Progress in all aspects of autonomous vehicles continues. A number of autonomous vehicle technologies appear in regular cars (albeit in the higher end models). Some worth mentioning are:

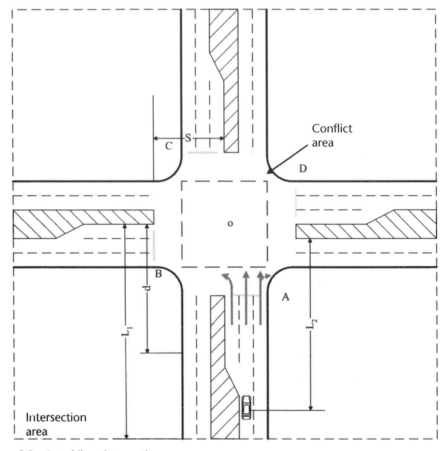

Figure 8.3 A multilane intersection.

- Lane departure warning systems;
- Lane departure correction systems;
- Advanced cruise control;
- Bus systems with precision docking;
- Collision warning (and safety precaution) systems;
- Dead-zone viewing cameras and warning systems;
- Active rearview mirrors.

During the year this book was being written and compiled, there were two interesting demonstrations of advanced technologies:

1. An Italian research group from the University of Parma drove a convoy of three semiautonomous vans from Rome to Shanghai (through Siberia) (http://viac.vislab.it/). Figure 8.5 shows one of the vans.
2. A group from Google developed and demonstrated a self-driving car in city traffic in U.S. West Coast cities with a log over 140,000 miles starting from San Francisco all the way around Lake Tahoe, indeed a very

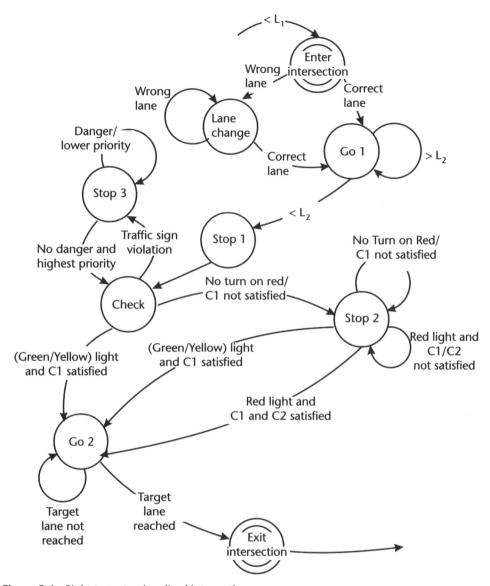

Figure 8.4 Right turn at a signalized intersection

noteworthy accomplishment (http://googleblog.blogspot.com/2010/10/what-were-driving-at.html).

Various demonstrations illustrating different aspects of autonomous driving are still continuing. The Grand Cooperative Driving Challenge, demonstrating semiautonomous driving with V2V and V2I communication, was held in the Netherlands (http://www.gcdc.net/).

An ITS-Energy demonstration, with three autonomous trucks, shown in Figure 8.6 with a very short following distance, will be showcased in Japan (http://www.fast-zero11.info/doc/Technical_Visits2.pdf).

Figure 8.5   Vans from VisLab. (Photo courtesy of Luciano Covolo.)

Figure 8.6   Three trucks demonstrating energy savings due to tight convoying.

References

[1]   Schneider, S., and Ü. Özgüner, "A Sensor Fault Controller Scheme to Achieve High Measurement Fidelity for Intelligent Vehicles, with Applications to Headway Maintenance," *Proceedings of the Intelligent Transportation Society of America 8th Annual Meeting and Exposition*, Detroit, May 1998, pp. 1–13.

[2]   Schneider, S., and Ü. Özgüner, "A Framework for Data Validation and Fusion, and Fault Detection and Isolation for Intelligent Vehicle Systems," *Proceedings of IV*, Stuttgart, Germany, October 1998.

[3]   Lee, S. C., "Sensor Value Validation Based on Systematic Exploration of the Sensor Redundancy for Fault Diagnosis," *IEEE Transactions on Systems, Man, and Cybernetics*, Vol. 24, April 1994, No. 4, pp. 594–605.

[4]   Liu, Y., and Ü. Özgüner, "Human Driver Model and Driver Decision Making for Intersection Driving," *Proc. IV'2007*, Istanbul, Turkey, pp. 642–647.

[5]   Acarman, T., et al., "Test-Bed Formation for Human Driver Model Development and Decision Making," *Proc. IEEE ITSC 2007*, Seattle, WA, 2007, -pp. 934–939.

[6]   Kurt, A., et al., "Hybrid-State Driver/Vehicle Modelling, Estimation and Prediction," *Proc. of 2010 13th International IEEE Conference on Intelligent Transportation Systems (ITSC)*, Madeira, Portugal, September 2010, pp. 806–811.

[7]   Tattegrain-Vest, H., et al., "Computational Driver Model in Transport Engineering: COSMODRIVE," *Journal of the Transportation Research Board*, Vol. 1550, 1996, pp. 1–7.

[8]   Delorme, D., and B. Song, *Human Driver Model for SMARTAHS*, Tech. Rep. UCB-ITS-PRR-2001-12, California PATH Program, Institute of Transportation Studies, University of California, Berkeley, April 2001.

[9]   Burnham, G., J. Seo, and G. Bekey, "Identification of Human Driver Models in Car Following," *IEEE Transactions on Automatic Control*, Vol. AC-19, December 1974, pp. 911–916.

[10]   Song, B., D. Delorme, and J. VanderWerf, "Cognitive and Hybrid Model of Human Driver," *2000 IEEE Intelligent Transportation Systems Conference Proceedings*, Dearborn, MI, October 2000, pp. 1–6.

[11]   Cody, D., S. Tan, and A. Garcia, *Human Driver Model Development*, Tech. Rep. UCB-ITS-PRR-2005-21, California PATH Program, Institute of Transportation Studies, University of California, Berkeley, June 2005.

Appendix

A.1 Two-Wheel Vehicle (Bicycle) Model

A single track vehicle or bicycle model is used to represent the simplified lateral dynamics of the vehicle model. The model covers two degrees of freedom along the following variables: lateral velocity and the angular velocity around the vertical axis, which is also called yaw rate. Steer angle and the longitudinal force on the front tire of the model are two input to the dynamical system.

The local coordinate system is fixed to the sprung mass CGs of the bicycle (single track) vehicle to describe the orientation and derive the motion dynamics of the vehicle mass. The global coordinate system fixed to the road is also defined to perceive rotation and displacement of the vehicle model from a stationary point. The longitudinal force on the front tire is denoted by $\frac{T_f}{R_f}$ where T_f is the traction or braking torque applied to the front axle and R_f is the effective radius of the tire. δ is the steer angle of the front tire. The forces generated in the lateral direction during turning are attached to the individual front and rear tire, denoted by F_{yf} and F_{yr}, respectively. The physical meanings are presented in Table A.1.

The motion dynamics in the lateral direction and the rotational dynamics around the local vertical axis is derived by summing the forces and the moments using the single track free body diagram given in Figure A.1. The reference direction of the states, forces, and front steer angle is chosen as illustrated in the top view of the bicycle model. The motion dynamics are derived with respect to these given reference directions.

The two-degree of freedom (2DOF) bicycle model is derived to represent the simplified lateral motion dynamics by ignoring all possible tire modeling complexities and other possible degrees of freedom on the rotational roll, pitch axes, and translational vertical and longitudinal axes. To maintain validity of the derived model, the operating point must be in the linear region of the tire force model characteristics and the small steer angle and the constant longitudinal velocity assumptions must hold (i.e., this model may be valid in the driving scenario when the vehicle is being driven at constant speed and low steering input is applied):

Table A.1   List of Physical Meanings of Nonlinear Bicycle Model

u	Longitudinal velocity of the vehicle model
v	Lateral velocity of the vehicle model
r	Yaw rate of the vehicle model
δ	Steering angle applied to the front wheel
a	Distance from the CG to the front axle
b	Distance from the CG to the rear axle
CG	Center of gravity of the vehicle
I_z	Rotational inertia in the horizontal plane
m	Total sprung and unsprung mass of the car
R_f	Effective rolling radius of the tire
k_f	Front tire lateral stiffness coefficient
k_r	Rear tire lateral stiffness coefficient
T_f	The drive torque on the front tire

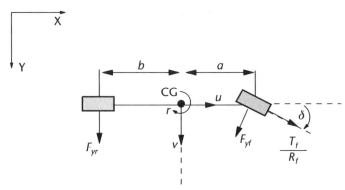

Figure A.1   Bicycle model of the car.

- The steering wheel angle is small and it is approximated as: $\sin\delta \approx \delta$ and $\cos\delta \approx 1$.
- The longitudinal velocity of the model is constant: $u = u_{constant}$.

Taking the sum of the forces in the lateral direction, which is shown by Y,

$$m\left(\dot{v} + ur\right) = F_{yf} + F_{yr} + \frac{T_f}{R_f}\delta$$

summing moments about the local Z-axis, which is perpendicular to the plane of the model:

$$I_z\dot{r} = aF_{yf} - bF_{yr} + a\frac{T_f}{R_f}\delta$$

In the linear and normal operating region of the tire characteristics, the tire force dynamic equations in the lateral direction can be linearized at the steady state and approximated as:

$$F_{yf} = k_f \alpha_f$$
$$F_{yr} = k_r \alpha_r$$

where α_f and α_r denote the front and rear tire slip angles, respectively. The tire slip angle is simply defined as the angular difference between the tire direction of motion and its orientation. For the determination of the slip angle on the front and rear tire we will be using,

$$\alpha_f = \delta - \tan^{-1}\left(\frac{v + ar}{u}\right)$$

$$\alpha_r = -\tan^{-1}\left(\frac{v - br}{u}\right)$$

The small slip angle values can be approximated as follows:

$$\alpha_f = \delta - \frac{v + ar}{u}$$

$$\alpha_r = -\frac{v - br}{u}$$

Inserting the lateral forces in the lateral dynamics of the bicycle model, the motion dynamics can be obtained by:

$$\dot{v} = -ur + \frac{1}{m}\left(k_f\left(\delta - \frac{v + ar}{u}\right) + k_r\left(-\frac{v - br}{u}\right) + \frac{T_f}{R_f}\delta\right)$$

$$\dot{r} = \frac{1}{I_z}\left(ak_f\left(\delta - \frac{v + ar}{u}\right) - bk_r\left(-\frac{v - br}{u}\right) + a\delta\frac{T_f}{R_f}\right)$$

From these equations, the linearized and simplified lateral dynamics are:

$$\dot{v} = -\left(\frac{k_f + k_r}{mu}\right)v - \left(u + \frac{ak_f - bk_r}{mu}\right)r + \frac{1}{m}\left(k_f + \frac{T_f}{R_f}\right)\delta$$

$$\dot{r} = -\frac{ak_f - bk_r}{I_z u}v - \frac{a^2 k_f + b^2 k_r}{I_z u}r + \frac{a}{I_z}\left(k_f + \frac{T_f}{R_f}\right)\delta$$

Hence, the linear bicycle model can be written in the state-space form:

$$\dot{x} = Ax + Bu$$

where the state vector is constituted by the lateral velocity and yaw rate $x = [v \quad r]^T$ and the input scalar is the steering wheel angle, $u = \delta$.

$$\begin{bmatrix} \dot{v} \\ \dot{r} \end{bmatrix} = \begin{bmatrix} -\left(\dfrac{k_f + k_r}{mu} \right) & -\left(u + \dfrac{ak_f - bk_r}{mu} \right) \\ \dfrac{ak_f - bk_r}{I_z u} & -\dfrac{a^2 k_f + b^2 k_r}{I_z u} \end{bmatrix} \begin{bmatrix} v \\ r \end{bmatrix} + \begin{bmatrix} \dfrac{1}{m}\left(k_f + \dfrac{T_f}{R_f} \right) \\ \dfrac{a}{I_z}\left(k_f + \dfrac{T_f}{R_f} \right) \end{bmatrix} \delta$$

A.2 Full Vehicle Model Without Engine Dynamics

In this section, the three-dimensional (3-D) nonlinear vehicle model is presented. The model is developed with the sprung mass of the vehicle having 6 degrees of freedom (6DOF) along the following variables: longitudinal velocity, lateral velocity, vertical velocity, yaw, roll, and pitch angular rates. The vehicle body is lifted by suspension systems at the front and rear axles. The rigid vehicle body (also called a *sprung mass*) is modeled as a lumped mass and moment of inertia at the center of gravity of the vehicle. The other masses such as wheels, brakes, and axles are referred to as *unsprung masses*. The local coordinate system is fixed to the sprung mass CGs of the double track model to describe the orientation and derive the motion dynamics of the vehicle mass.

A.2.1 Lateral, Longitudinal, and Yaw Dynamics

The double track vehicle model represents the physical model of the vehicle. The top view of the vehicle model is shown in Figure A.2 where u represents the longitudinal velocity, v is the lateral velocity, and r is the yaw rate. F_{xi} and F_{yi} represent

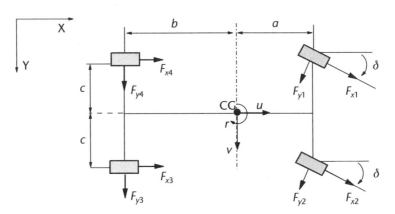

Figure A.2 Top view of the vehicle.

the individual longitudinal and lateral force between the tire patches and the road for $i = 1, 2, 3, 4$, respectively.

The distance from the tires to the sprung mass center of gravity (CG) is denoted by c, and the distances from the front and rear axle to CG are represented by a and b, respectively. The steer angle of the front tires is represented by δ, whereas the rear tires are assumed to have a zero steer angle. The summation of the forces in the X-direction yields acceleration in the longitudinal direction,

$$\dot{u} = vr + \frac{1}{m}\left(\left(F_{x1} + F_{x2}\right)\cos\delta + F_{x3} + F_{x4} - \left(F_{y1} + F_{y2}\right)\sin\delta - A_\rho u^2 sign(u)\right) + g\sin(\theta_T)$$

where m is the total mass of the vehicle including unsprung masses, A_ρ is the aerodynamic drag coefficient, and θ_T is the terrain angle affecting the longitudinal dynamics.

Acceleration in the lateral direction is derived by summing the forces in the Y direction,

$$\dot{v} = -ur + \frac{1}{m}\left(\left(F_{y1} + F_{y2}\right)\cos\delta + F_{y3} + F_{y4} + \left(F_{x1} + F_{x2}\right)\sin\delta\right) + g\cos(\theta_T)\sin(\Phi_T)$$

where Φ_T is the terrain angle as illustrated in Figure A.3.

Totaling moments about the local Z-axis of the vehicle model gives the time-derivative of yaw rate:

$$\dot{r} = \frac{1}{I_z}\left[-b\left(F_{y3} + F_{y4}\right) + c\left(F_{x4} - F_{x3}\right) + a\left(\left(F_{x1} + F_{x2}\right)\sin\delta + \left(F_{y1} + F_{y2}\right)\cos\delta\right)\right.$$
$$\left. + c\left(\left(F_{y2} - F_{y1}\right)\sin\delta + \left(F_{x1} - F_{x2}\right)\cos\delta\right)\right]$$

where I_z is the moment of inertia of the total mass of the vehicle about the local Z-axis.

The front view of the 3-D vehicle model is illustrated in Figure A.4. The sprung mass (body of the vehicle) is connected to the unsprung masses (front and rear axles and the individual tires) through the rocker arms. Individual suspension forces

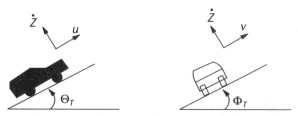

Figure A.3   Road inclination and terrain angles affecting the vehicle dynamics.

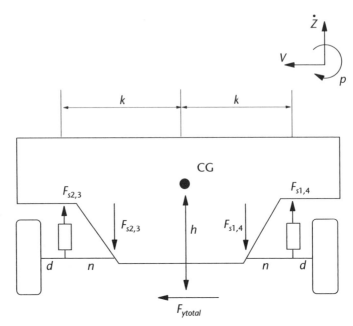

Figure A.4   Front view of the vehicle.

are denoted by F_{si} for $i = 1, 2, 3, 4$ and the suspension mechanisms are drawn as rectangular boxes for simplicity.

The sprung mass CG height of the vehicle versus the ground plane reference is given by h. The mechanical link between the wheel and the sprung mass is supported by the rocker arm and suspension mechanism. The distance between the contact point of the suspension rocker and the wheel is denoted by d, whereas the distance between the contact point of the suspension rocker and the body is denoted by n. The distance between the contact point of the suspension rocker and the CG of the vehicle model is denoted by k.

The summation for the forces about the local Z-axis gives the acceleration in the vertical direction, which is denoted by $\ddot{Z}$,

$$\ddot{Z} = uq - vp + \frac{1}{m_s} \sum_{i=1}^{4} \frac{F_{si}}{R_{si}} - g\cos(\theta_T)\cos(\phi_T)$$

where m_s is the mass of the sprung mass of the vehicle model. Roll dynamics are obtained by summing the moment about the local X-axis,

$$\dot{p} = \frac{1}{I_{xs}} \left(\sum_{i=1}^{4} \overline{R}_{ri} F_{si} + F_{ytotal} h + qr(I_{ys} - I_{zs}) \right)$$

where I_{xs}, I_{ys}, and I_{zs} are the moments of inertias of the sprung mass of the vehicle about the local X-, Y-, and Z-axes, respectively, and $\overline{R}_{ri}$ for $i = 1, 2, 3, 4$ is given in terms of the rocker arm displacement versus the wheel contact point, sprung mass contact point, and CG of the vehicle model:

$$\bar{R}_{r1} = -\bar{R}_{r2} = -\bar{R}_{r3} = \bar{R}_{r4} = -k + (k - n)\frac{d}{d + n}$$

Taking the summation of the moments about the Y-axis gives the pitch dynamics,

$$\dot{q} = \frac{1}{I_{ys}}\left(F_{xtotal}h + a\left(\frac{F_{s1}}{R_{r1}} + \frac{F_{s2}}{R_{r2}}\right) - b\left(\frac{F_{s3}}{R_{r3}} + \frac{F_{s4}}{R_{r4}}\right) + pr\left(I_{zs} - I_{xs}\right)\right)$$

where the sum of the forces in the longitudinal direction is given by

$$F_{xtotal} = \left(F_{xq} + F_{x2}\right)\cos\delta + F_{x3} + F_{x4} + \left(F_{yq} + F_{y2}\right)\sin\delta$$

and the sum of the forces in the lateral direction is given by

$$F_{ytotal} = \left(F_{yq} + F_{y2}\right)\cos\delta + F_{y3} + F_{y4} - \left(F_{x1} + F_{x2}\right)\sin\delta$$

A.2.2   Suspension Forces and Tire Dynamics

A passive, spring damper model is used for the suspension actuators as shown in Figure A.5. The suspension force generated in the vertical direction, denoted by F_{si}, is given by:

$$F_{si} = K_{si}\left(Z_{ui} - Z_{si} + L_{si}\right) + C_{si}\left(\dot{Z}_{ui} - \dot{Z}_{si}\right)$$

for $i = 1, 2, 3, 4$ where Z_{si} and Z_{ui} are the heights of the ith corner and the vertical height of the ith unsprung mass above the ground reference, respectively, L_{si} is the unweighted length of the ith suspension actuator, K_{si} is the spring constant, and C_{si} is the damper coefficient of the ith suspension actuator.

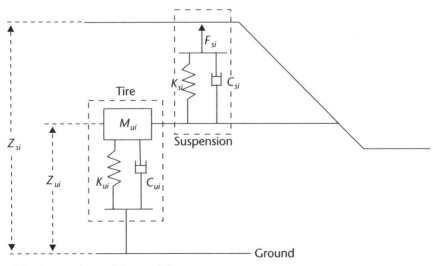

Figure A.5   Suspension and tire models.

Similarly, the unsprung masses are modeled as a spring and a damper in parallel and Z_{ui} satisfy the following differential equation:

$$m_{ui}\ddot{Z}_{ui} = K_{ui}\left(R_{fi} - Z_{ui}\right) - C_{ui}\dot{Z}_{ui} - F_{wi} - gm_{ui}$$

where R_{fi} is the radius of the ith unloaded tire, K_{ui} and C_{ui} are the spring and the damper constant of the ith tire model, and m_{ui} is the unsprung mass.

The corner heights of sprung masses need to be calculated in terms of the known quantities. The heights of each corner at any time instant can be found by adding the displacements due to roll and pitch to the initial height as follows:

$$Z_{s1} = Z_{s1ss} + Z_s - k\sin(\phi_T) + a\sin(\theta_T)$$
$$Z_{s2} = Z_{s2ss} + Z_s + k\sin(\phi_T) + a\sin(\theta_T)$$
$$Z_{s3} = Z_{s3ss} + Z_s + k\sin(\phi_T) - b\sin(\theta_T)$$
$$Z_{s4} = Z_{s4ss} + Z_s - k\sin(\phi_T) - b\sin(\theta_T)$$

where Z_s is the vertical displacement of the CG and Z_{siss} is the steady-state height of the ith corner of the related sprung mass.

A.2.3   Tire Forces

Towards 3-D nonlinear vehicle model derivation, a nonlinear tire model capable of presenting possible saturation effects is used to model individual tire forces along the longitudinal and lateral axes. A semiempirical tire model, also referred as the *magic formula*, is introduced by Pacejka. Dugoff is an alternative analytical tire model, which illustrates the tire force characteristics versus slip and other states. Combined longitudinal and lateral force generation are directly related to the individual tire road coefficient in compact form, for $i = 1, 2, 3, 4$.

$$F_{xi} = k_{xi}\frac{\lambda_i}{1+\lambda_i}f(\gamma_i)$$
$$F_{yi} = k_{yi}\frac{\tan(\alpha_i)}{1+\lambda_i}f(\gamma_i)$$

where k_{xi} and k_{yi} are the tire longitudinal and lateral cornering stiffness, respectively. The variable γ_i is expressed in term of the individual slip ratio, λ_i, the slip angle α_i, the tire-road friction coefficient, denoted by μ, and the vertical force on the tire, F_{zi}, as follows:

$$\gamma_i = \frac{\mu F_{zi}\left(1+\lambda_i\right)}{2\sqrt{\left(k_{xi}\lambda_i\right)^2 + \left(k_{yi}\tan(\alpha_i)\right)^2}}$$

The function $f(\gamma_i)$ is given by

$$f(\gamma_i) = \begin{cases} (2-\gamma_i)\gamma_i & \text{if} \quad \gamma_i < 1 \\ 1 & \text{if} \quad \gamma_i \geq 1 \end{cases}$$

The individual tire slip angles are denoted by α_i,

$$\alpha_1 = \delta - \tan^{-1}\left(\frac{v+ar}{u+cr}\right)$$

$$\alpha_2 = \delta - \tan^{-1}\left(\frac{v+ar}{u-cr}\right)$$

$$\alpha_3 = -\tan^{-1}\left(\frac{v-br}{u-cr}\right)$$

$$\alpha_4 = -\tan^{-1}\left(\frac{v-br}{u+cr}\right)$$

The slip ratio for each individual tire is given as

$$\lambda_i = \begin{cases} \dfrac{Rw_i - u_{ti}}{Rw_i} & \text{if} \quad Rw_i \geq u_{ti} \quad \text{(during acceleration)} \\[2ex] \dfrac{Rw_i - u_{ti}}{u_{ti}} & \text{if} \quad Rw_i < u_{ti} \quad \text{(during braking)} \end{cases}$$

where R is the tire effective radius and u_{ti} denotes the velocity on the rolling direction from the ith individual tire model,

$$u_{t1} = (\;+\;)\cos\delta + (\;+\;)\sin\delta$$
$$u_{t2} = (\;-\;)\cos\delta + (\;+\;)\sin\delta$$
$$u_{t3} = u - cr$$
$$u_{t4} = u + cr$$

The vertical tire force is given subject to dynamic weight transfer excited by pitch and roll motions,

$$F_{z1} = \frac{a}{2l}mg - \frac{mh_{cg}}{l}\dot{u} + \frac{amh_{cg}}{2el}\dot{v} + eC_{s1}\dot{p} + eK_{s1}p$$

$$F_{z2} = \frac{a}{2l}mg - \frac{mh_{cg}}{l}\dot{u} + \frac{amh_{cg}}{2el}\dot{v} - eC_{s2}\dot{p} - eK_{s2}p$$

$$F_{z3} = \frac{a}{2l}mg + \frac{mh_{cg}}{l}\dot{u} - \frac{amh_{cg}}{2el}\dot{v} + eC_{s3}\dot{p} + eK_{s3}p$$

$$F_{z4} = \frac{a}{2l}mg + \frac{mh_{cg}}{l}\dot{u} - \frac{amh_{cg}}{2el}\dot{v} - eC_{s4}\dot{p} - eK_{s4}p$$

where $l = a + b$ is the wheel base, p, $\dot{p}$ denotes the roll angle and its time derivative, and C_{si} and K_{si} are the individual suspension damping and stiffness coefficients, respectively. The variables and parameters used in the modeling study are denoted in Table A.2.

Nonlinear characteristics' response of the combined longitudinal and lateral tire force model with respect to the slip ratio and the side-slip angle variable change is simulated in Figures A.6 and A.7, respectively.

Table A.2 List of Physical Meanings of the 3-D Nonlinear Vehicle Model

u	Longitudinal velocity of the vehicle	v	Lateral velocity of the vehicle
$\dot{Z}$	Vertical velocity of the vehicle	r	Yaw rate of the vehicle
p	Roll rate of the vehicle	q	Pitch rate of the vehicle
m	The total mass of the vehicle	m_s	The sprung mass of the vehicle
I_x	The moment of inertia about the local X-axis	I_{xs}	The moment of inertia of the sprung mass about the local X-axis
I_y	The moment of inertia about the local Y-axis	I_{ys}	The moment of inertia of the sprung mass about the local Y-axis
I_z	The moment of inertia about the local Z-axis	I_{zs}	The moment of inertia of the sprung mass about the local Z-axis
F_{xi}	The longitudinal force between tires and road	F_{yi}	The lateral force between tires and road
δ	Front steering angle	m_{ui}	The unsprung mass
α_i	The individual tire slip angle	λ_i	The slip ratio for each individual tire
F_{si}	Suspension forces	Z_{ui}	The height of the ith unsprung mass
Z_{si}	The height of the ith corner of the related sprung	K_{ui}	The spring constant of the ith tire model
K_{si}	The spring constant of the ith suspension actuator	C_{ui}	The damper constant of the ith tire model
C_{si}	The damper constant of the ith suspension actuator	$\bar{R}_{ri}$	The equilibrium of the rocker in the Z direction
R_{fi}	The radius of the ith unloaded tire	F_{zi}	The vertical force exerted on the ith unsprung mass
ϕ_T	Terrain angle	θ_T	Terrain angle
d	The distance between the rocker arm and the wheel	n	The distance between the contact point of the suspension rocker arm and the body
A_ρ	Aerodynamic drag coefficient	k	The distance between the rocker arm and the CG of the vehicle model

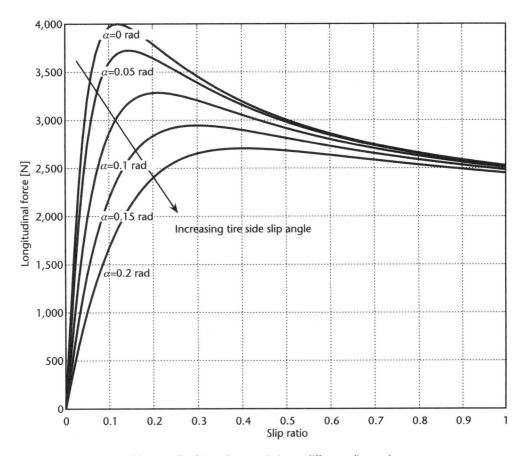

Figure A.6   Combined longitudinal tire characteristics at different slip angles.

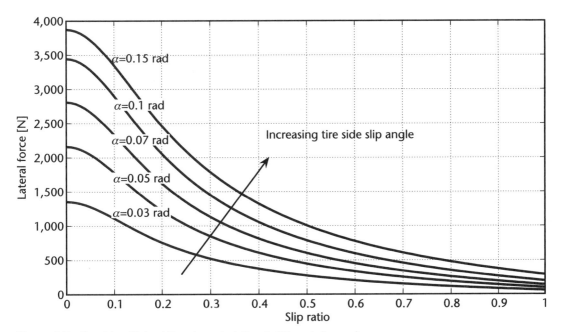

Figure A.7   Combined lateral tire characteristics at different slip angles.

About the Authors

Ümit Özgüner received a Ph.D. from the University of Illinois and held positions at IBM, the University of Toronto, and Istanbul Technical University. He is a professor of electrical and computer engineering and holds the TRC Inc. chair on intelligent transportation systems (ITS) at The Ohio State University. He is a fellow of the IEEE. Professor Özgüner's areas of research interest are in ITS, decentralized control, and autonomy in large systems, and is the author of over 400 publications. He was the first president of the IEEE ITS Council in 1999 as it transformed into the IEEE ITS Society and he has also been the ITS Society vice president for conferences. Professor Özgüner has also served the IEEE Control Society in various positions. He participated in the organization of many conferences and was the program chair of the first IEEE ITS Conference and the general chair of the IEEE Control Systems Society 2002 CDC, ITS Society IV 2003, and ICVES 2008. Teams coordinated by Professor Özgüner participated successfully in the 1997 Automated Highway System Technology Demonstration, the DARPA 2004 and 2005 Grand Challenges, and the 2007 Urban Challenge.

Tankut Acarman received a B.S. in electrical engineering and a master's degree in computer and control engineering from Istanbul Technical University, Istanbul, Turkey, in 1993 and 1996. He received a Ph.D. in electrical and computer engineering from The Ohio State University, Columbus, Ohio, in 2002. He has been an associate professor of computer engineering at Galatasaray University since 2004. He has been a vice director of the Institute of Sciences since 2008. Dr. Acarman was a faculty member in a potential center for a network of excellence funded by EU-FP6 to introduce actively safe, clean, and efficient road vehicles. He has served as an expert in various projects, including a national project aiming to reduce traffic accidents by improving driving and driver prudence through signal processing and advanced information technologies, and energy management strategies for hybrid electric vehicle technologies supported by R&D departments of the automotive manufacturers in Turkey. His research interests lie in all aspects of intelligent vehicle technologies, driver assistance systems, and performance evaluation of intervehicle communication in highway systems and in urban areas. He is currently contributing to a national project aiming to develop an intervehicle communication system funded by the Scientific & Technological Research Council of Turkey and coordinated by two automotive manufacturers. He is an associate editor of *IEEE*

Technology News covering intelligent transportation systems. He is also a member of the IEEE.

Keith Redmill received a B.S.E.E. and a B.A. in mathematics from Duke University in 1989 and an M.S. and a Ph.D. from The Ohio State University in 1991 and 1998, respectively. He has been with the Department of Electrical and Computer Engineering since 1999, initially as a senior research associate and most recently as a research scientist. Dr. Redmill has led or participated in a wide range of interdisciplinary projects including a series of self-driving automated passenger vehicles, autonomous ground and aerial robotic vehicle development and experiments, sensing and sensor fusion development projects involving computer vision, LADAR, radar, GPS, IMU, and other sensing modalities, wireless vehicle-to-vehicle and vehicle-to-infrastructure communication simulation and application development, traffic monitoring and data collection, intelligent vehicle control and safety systems for vehicles ranging from small ATVs to heavy duty commercial trucks, remote sensing programs, embedded and electromechanical system design and prototyping, and process control development. His areas of technical interest include control and systems theory, intelligent transportation systems, autonomous vehicle and robotic systems, real-time embedded systems, GPS and inertial positioning and navigation, transit and traffic monitoring, image processing, wireless digital communication for vehicles, sensor technologies, decentralized multiagent hierarchical and hybrid systems, and numerical analysis and scientific computing. He has extensive software development, electronics development and testing, and embedded systems deployment experience. He is a member of the IEEE and SIAM.

Index

For further information on these and other Artech House titles, including previously considered out-of-print books now available through our In-Print-Forever® (IPF®) program, contact:

Artech House
685 Canton Street
Norwood, MA 02062
Phone: 781-769-9750
Fax: 781-769-6334
e-mail: artech@artechhouse.com

Artech House
16 Sussex Street
London SW1V HRW UK
Phone: +44 (0)20 7596-8750
Fax: +44 (0)20 7630 0166
e-mail: artech-uk@artechhouse.com

Find us on the World Wide Web at: www.artechhouse.com